# SNOWFLAKES
## in Photographs

W. A. Bentley

DOVER PUBLICATIONS
GARDEN CITY, NEW YORK

*Bibliographical Note*

This Dover edition, first published in 2000, is a selection of 72 plates from the work published in 1962 by Dover Publications which was, in turn, an unabridged reprint of the book originally published in 1931 by the McGraw-Hill Book Company, Inc. under the title *Snow Crystals*, with photographs by W. A. Bentley and text by W. J. Humphreys. The Introduction is a reprint of the text of the article, "Photographing Snowflakes," by Wilson A. Bentley, which originally appeared in 1922 in *Popular Mechanics* magazine, Vol. 37, pages 309–312.

DOVER *Pictorial Archive* SERIES

*Library of Congress Cataloging-in-Publication Data*

Bentley, W. A. (Wilson Alwyn), 1865–1931.
 Snowflakes in photographs / W.A. Bentley.
  p. cm.
 Reprint. Originally published under title: Snow crystals. New York : McGraw-Hill, 1931.
  ISBN-13: 978-0-486-41253-5 (pbk.)
  ISBN-10: 0-486-41253-9 (pbk.)
  1. Snowflakes—Pictorial works. I. Bentley, W. A. (Wilson Alwyn), 1865–1931. Snow crystals. II. Title.

QC926.36 .B46 2000
551.57'841—dc21

                                                        00-038336

Manufactured in the United States of America
41253920   2026
www.doverpublications.com

# INTRODUCTION*

## Photographing Snowflakes
### by Wilson A. Bentley

Every snowflake has an infinite beauty which is enhanced by knowledge that the investigator will, in all probability, never find another exactly like it. Consequently, photographing these transient forms of Nature gives to the worker something of the spirit of a discoverer. Besides combining her greatest skill and artistry in the production of snowflakes, Nature generously fashions the most beautiful specimens on a very thin plane so that they are specially adapted for photomicrographical study.

The photographing of snowflakes, although quite delicate work, can hardly be called difficult, although some hardships attend it, because the work must all be done in a temperature below freezing, and under conditions of much physical exposure. The temperature at which photography is possible depends somewhat upon the thickness of the crystals; this varies greatly from time to time, and depends upon whether the temperature is rising from an intense degree of cold or falling from a point above freezing. If rising after a cold snap, photographing can often be continued until actual thawing commences.

Of course, location is everything in this work, and no one except those living in arctic climates or in regions having long and severe winters, can accomplish much. Generally speaking, the western quadrants of widespread storms or blizzards furnish the most beautiful and perfect forms. At such times the wind is usually westerly or northerly, with the barometer standing at 29.6 to 29.9 in. and slowing rising. The percentage of perfect crystals is likely to be larger when the snowfall is not too thick and heavy, with the crystals medium to small in size rather than large. The character of the snowfall often undergoes quite abrupt changes as a storm progresses.

The apparatus required for snowflake photography consists of a compound microscope, fitted with a joint that permits the instrument to be turned down horizontally, at right angles to its base, so that it can be coupled to a camera bellows by means of a light-tight connection. The microscope objectives are used alone, without the eyepiece. It is best to have several different objectives; 1/2, 3/4, and 3-in. combinations, which give magnifications of from 8 to 60 diameters (64 to 3,600 times), will serve well.

Ordinary daylight, coming through a window, is used for illuminating the crystal after it has been placed on a microscope slide, a tiny beam of light entering through the small aperture in the substage of the instrument. The apparatus is placed indoors, near by and facing a window. The room, the apparatus, and its accessories should always be away from any source of artificial heat, and at a temperature approximately that of the outside air. The necessary accessories are an observation microscope, a pair of thick mittens, microscope slides, a sharp-pointed wooden splint, a feather, and a turkey wing or similar duster; also, an extra focusing back for the camera, containing clear glass instead of the usual ground glass, with a magnifying lens attached; this is used for final focusing. A blackboard, about 1 ft. square, with stiff wire or metal handles at the ends, so that the hands will not touch and warm it, it used to collect the specimens. As it is necessary to cover the end of the microscope objective with a strip of black card, that takes the place of the usual camera shutter which controls the duration of exposure, it is necessary to fit two vertical rods at each side of the microscope tube to hold the card.

The snowflakes are caught on the blackboard as they fall, and examined by the naked eye or with

---

*This Introduction is a reprint of the text of an article originally published in 1922 in *Popular Mechanics* magazine, Vol. 37, pages 309–312. It is presented here in its entirety except for two *very* slight alterations, both indicated by ellipsis: one to the first sentence in the eighth paragraph and another to the last sentence of the final paragraph of the piece, both of which refer to photographs in the original that are not included here.

the assistance of a hand magnifying glass. The feather duster is used to brush the board clean every few seconds, until two or more promising specimens alight upon it, when it is immediately removed indoors. From this point onward the photographer must work fast. The promising specimens are placed for a moment's observation under the observation microscope. The removal of the snowflake from the board to the microscope slide is accomplished with the sharp-pointed splint, which is pressed gently against the face of the crystal until the latter adheres to it, so that it can be picked up and placed on the glass slide. Usually several crystals are placed together on a single slide, a momentary glance being given to each, and care taken while doing this not to breathe on the crystals. The utmost haste must be used, for a snow crystal is often exceedingly tiny, and frequently not thicker than heavy paper. Furthermore, once these bits of pure beauty are isolated, evaporation (not melting) soon wears them away, so that, even in zero weather, they last but a very few minutes. When a desirable specimen is obtained, it is pressed flat against the glass with the edge of the feather and the slide inserted in the stage of the microscope on the camera stand, centered, roughly focused with the camera ground glass, then sharply focused with the clear-glass screen and magnifier, focusing on some tiny air tube near the center of the crystal. The plate holder is then inserted into the camera, the objective covered with the black card and the slide removed from the plate holder. The objective is then uncovered, and when the exposure, which may vary from 8 seconds to 100 or more, is deemed sufficient, the operation is reversed. Naturally enough, no rule for the length of exposure can be given, except that the greater the magnification, the longer the exposure should be.

The frail, feathery flakes are the most difficult to photograph, and it is always best to place five or six other crystals around the specimen, as this greatly retards the evaporation of the central one.

When working from the rear of the camera, and the bellows extension is such as to make it impossible to reach the focusing screw on the microscope, an arrangement can be used . . . [that] . . . consists of a cord that runs over a wheel on each side of the camera and around the focusing screw. No lens is required in the camera, the microscope furnishing the optical equipment for projecting the images onto the sensitized plates.

Having recorded the fleeting substance of the snowflakes on the photographic negative and brought out the image by development, the photographer discovers that the body of the snow crystal is so transparent, that it does not contrast enough with its background to make a print in which the form will stand out in relief. There is no purely photographic method for producing the white images against a dark background, and yet it is necessary to do so if the images are to be appreciated by most people, whose ideal of snow is that of immaculate whiteness. The only effective method of accomplishing this result is what is known among photographers as "blocking out."

The negative is supported on an ordinary retoucher's desk, which may be merely a piece of glass, arranged to hold the negative so that the image is illuminated by transmitted light. Then, with an etching knife or other fine, sharp-pointed tool, the operator proceeds to scrape away the emulsion around the outline of the crystal to leave it standing alone against a background of clear glass. This requires considerable patience, and often considerable time as well. In order to avoid irreparably spoiling the original negative, it is best not to alter it in any way, but to make a copy negative on which the actual blocking out is done. After the negative has been thus prepared, prints or lantern slides are made in the usual manner. Blocking out the negatives is done indoors. . . .

# CONTENTS

Reference is by number and page. The numbers are continuous, line by line, across the page as one would read.

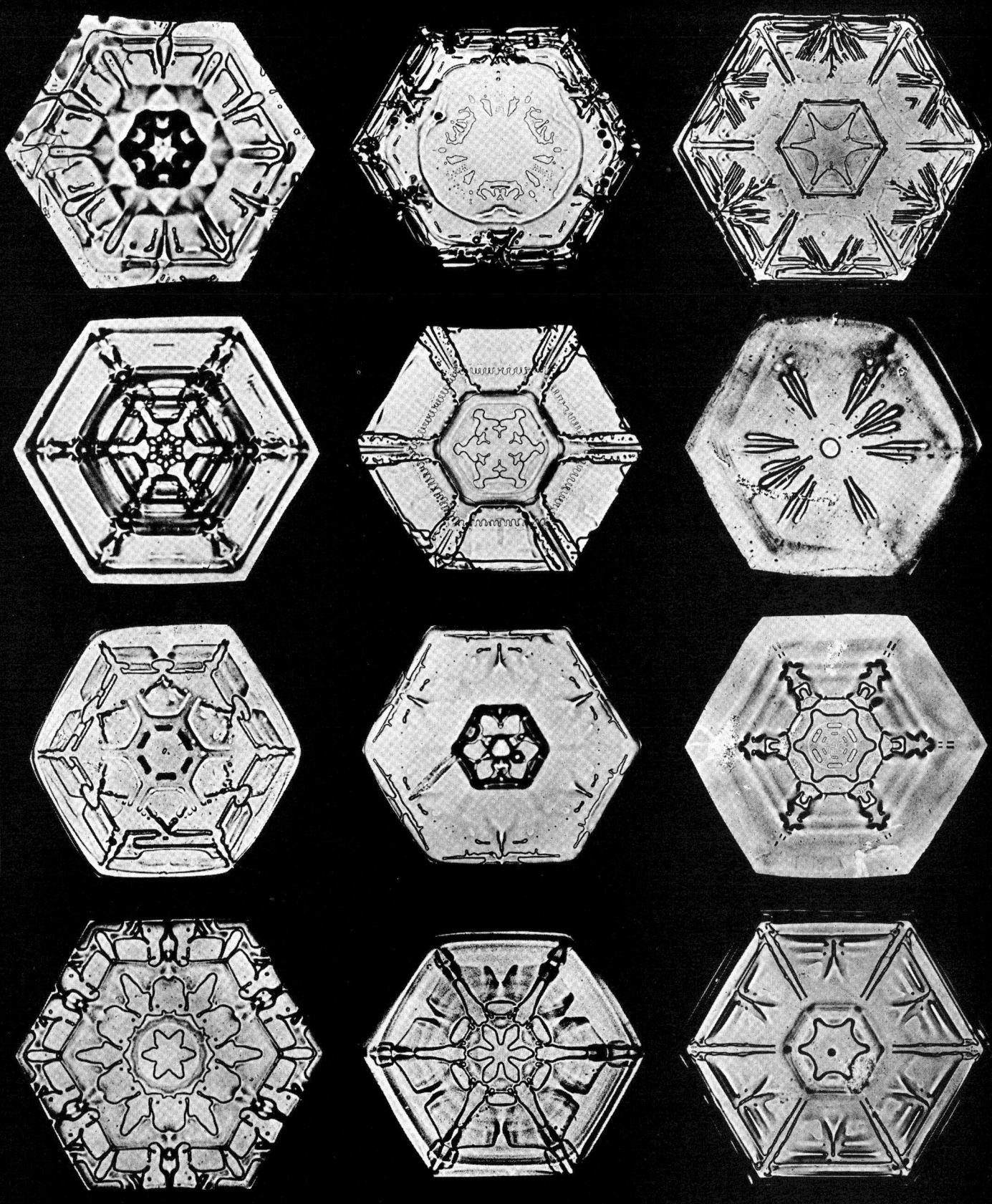

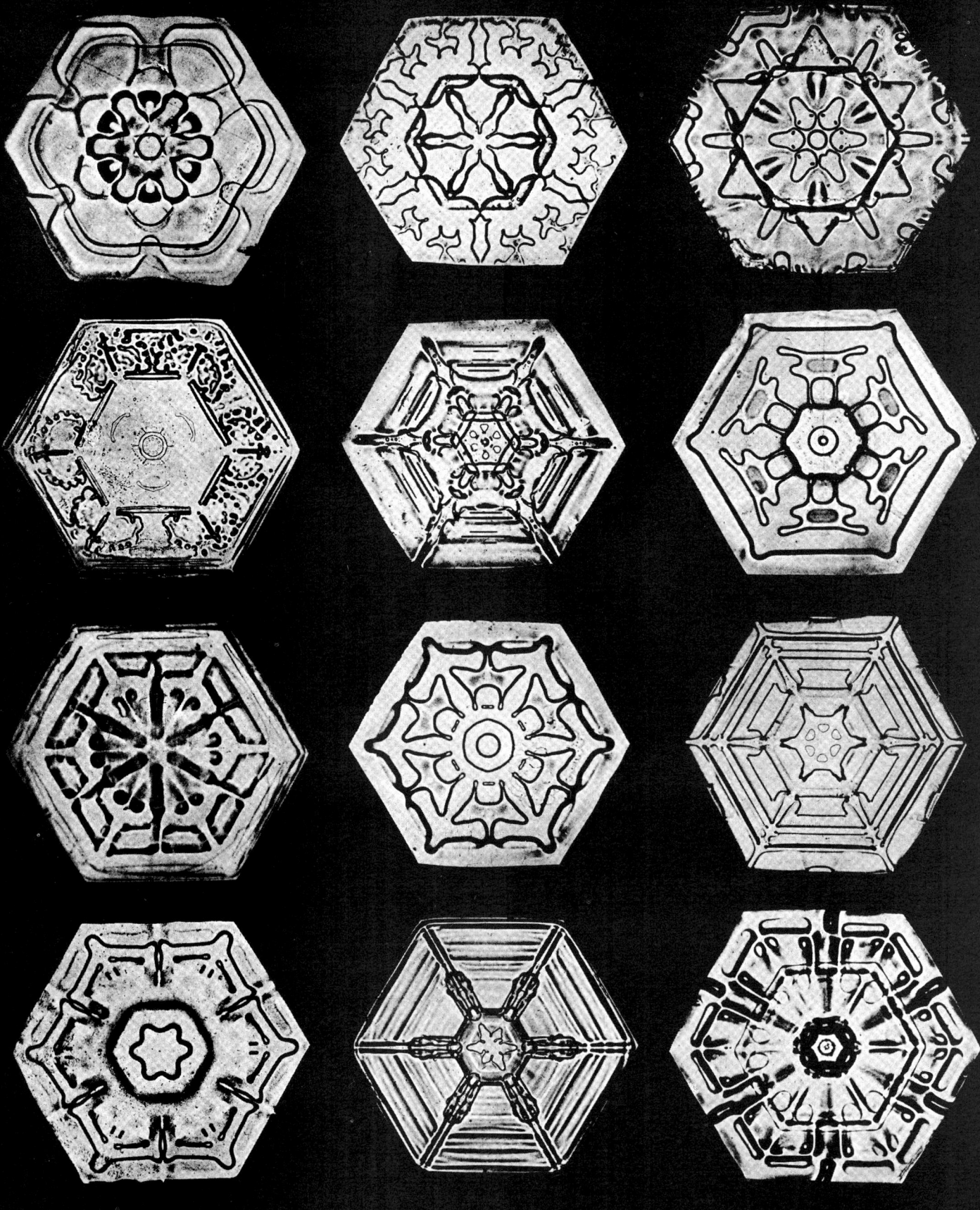

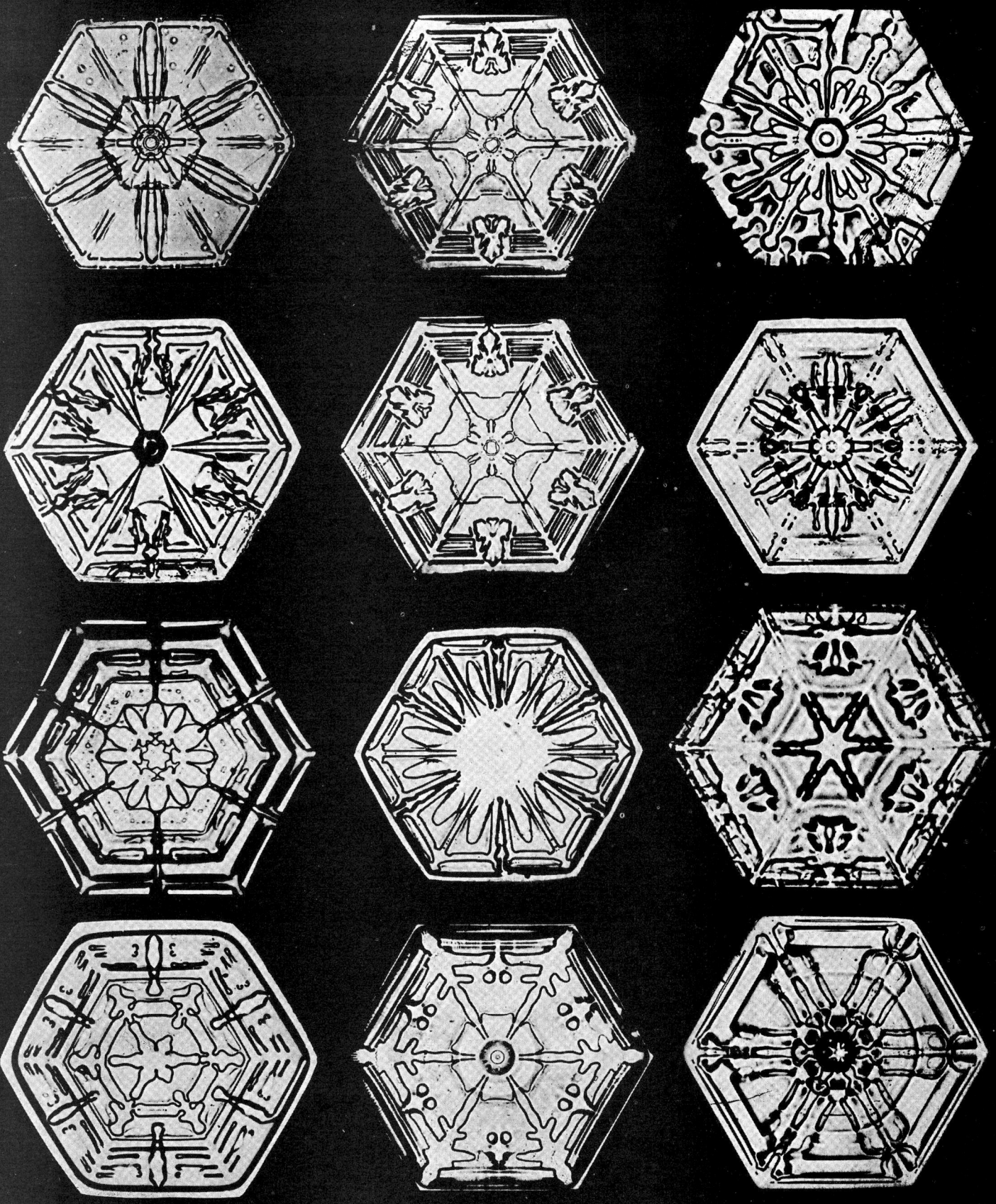

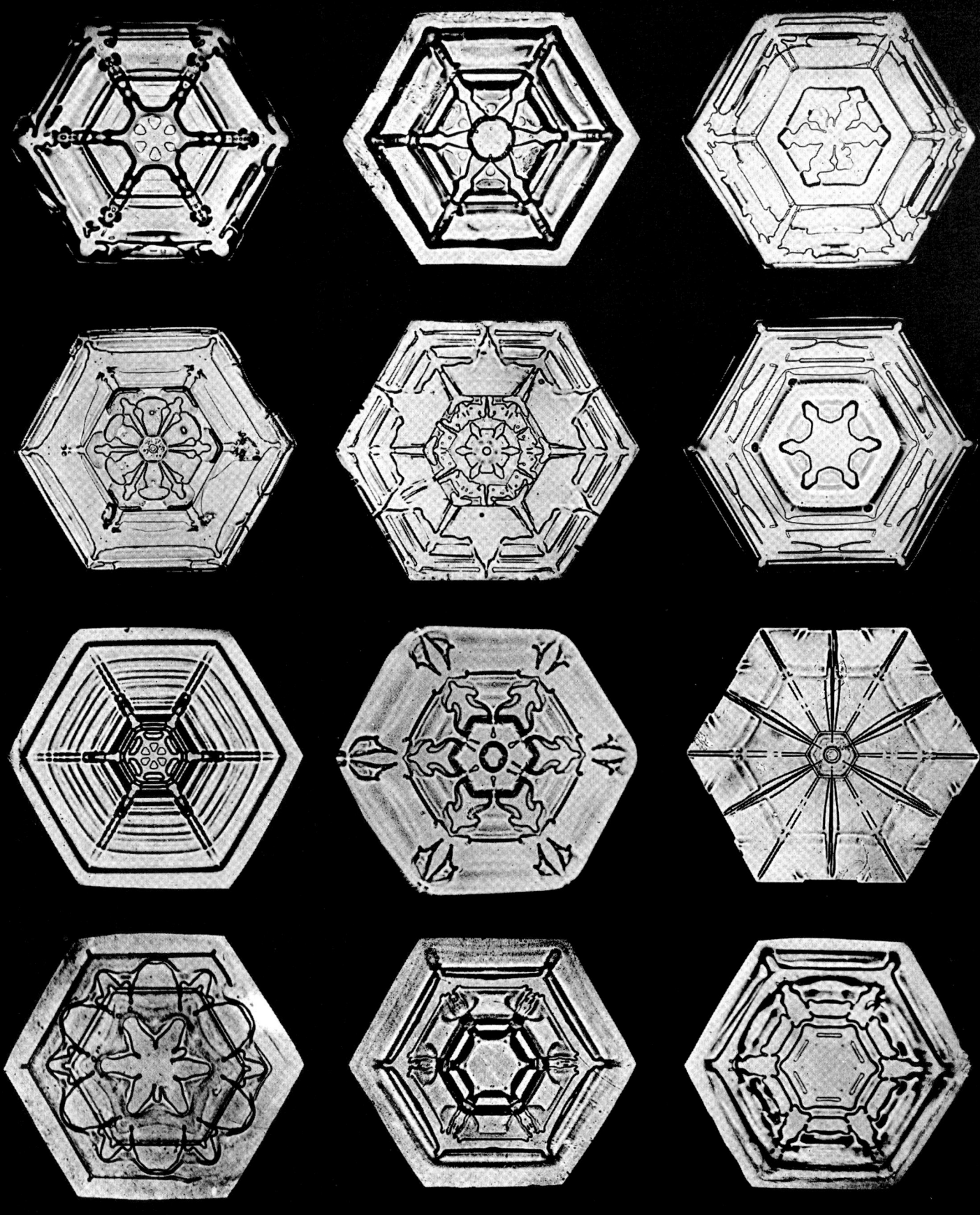

4

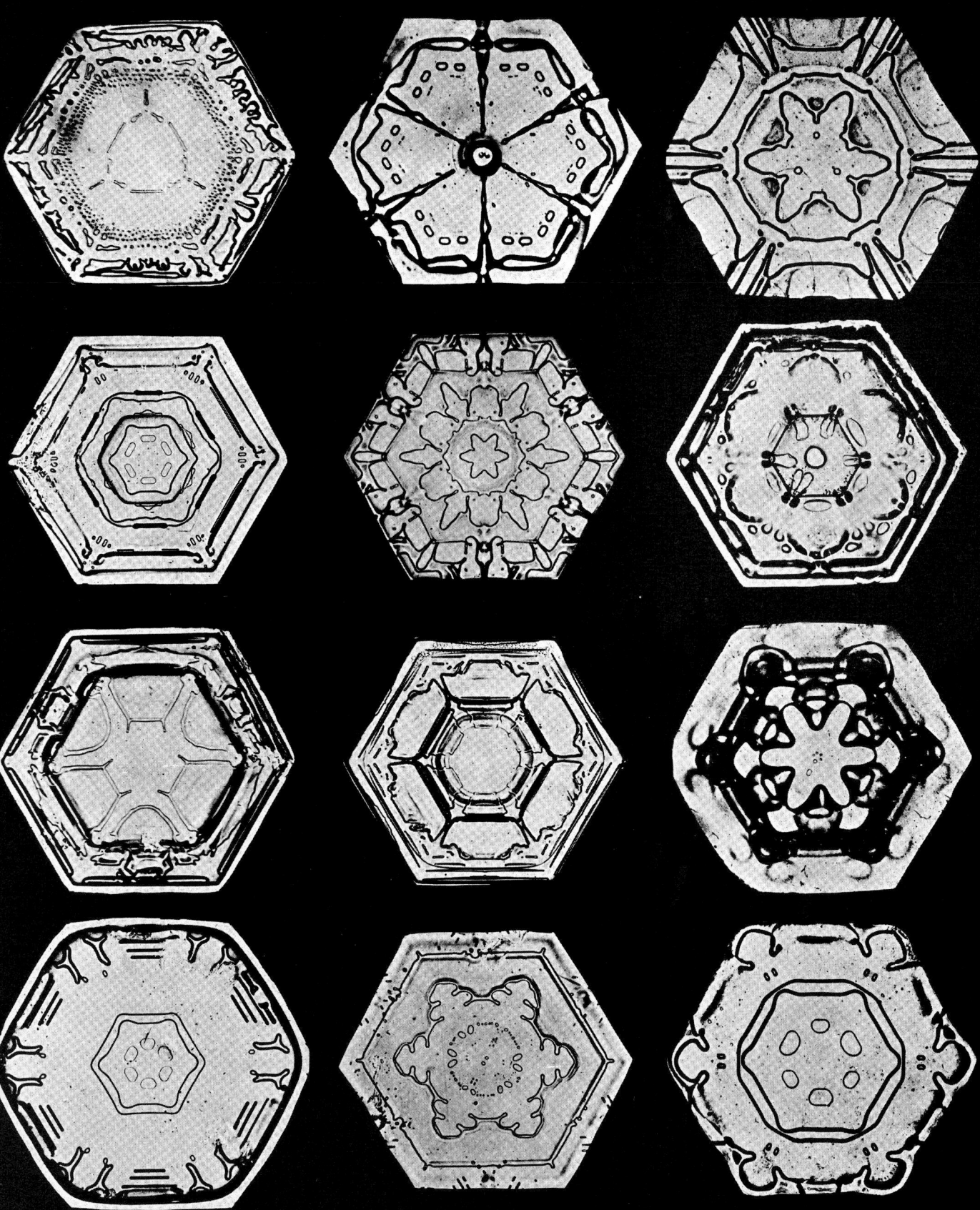

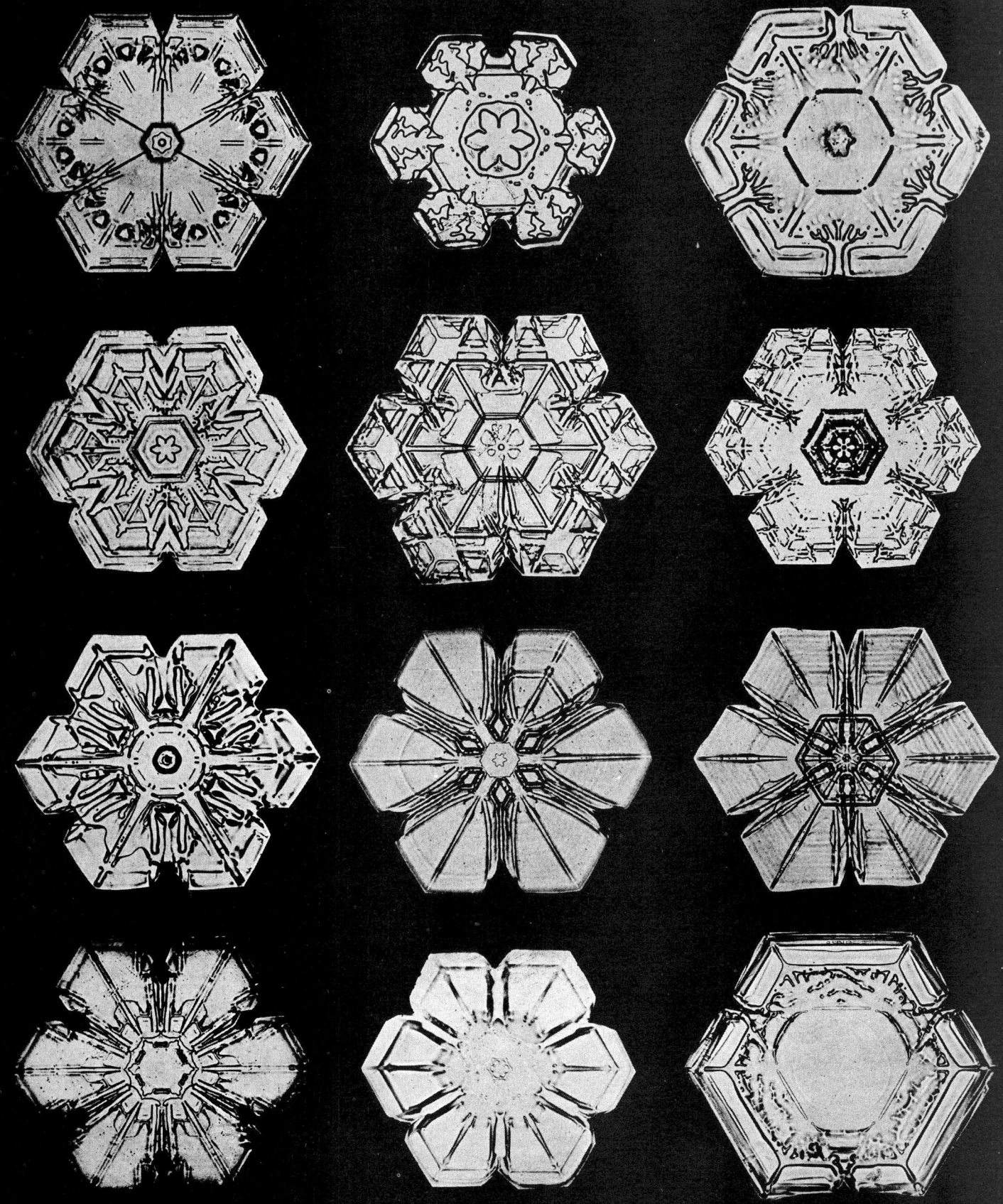

6

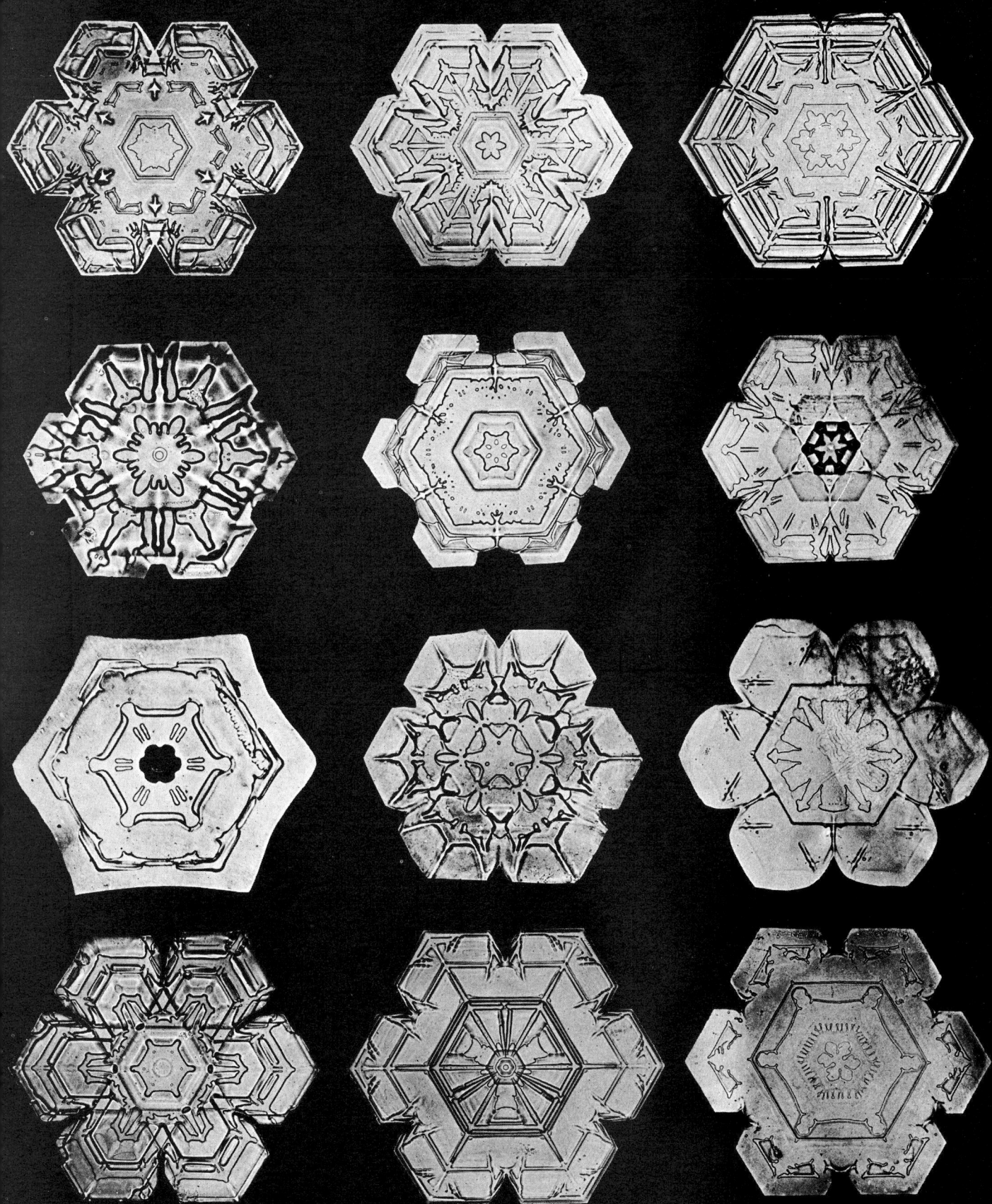

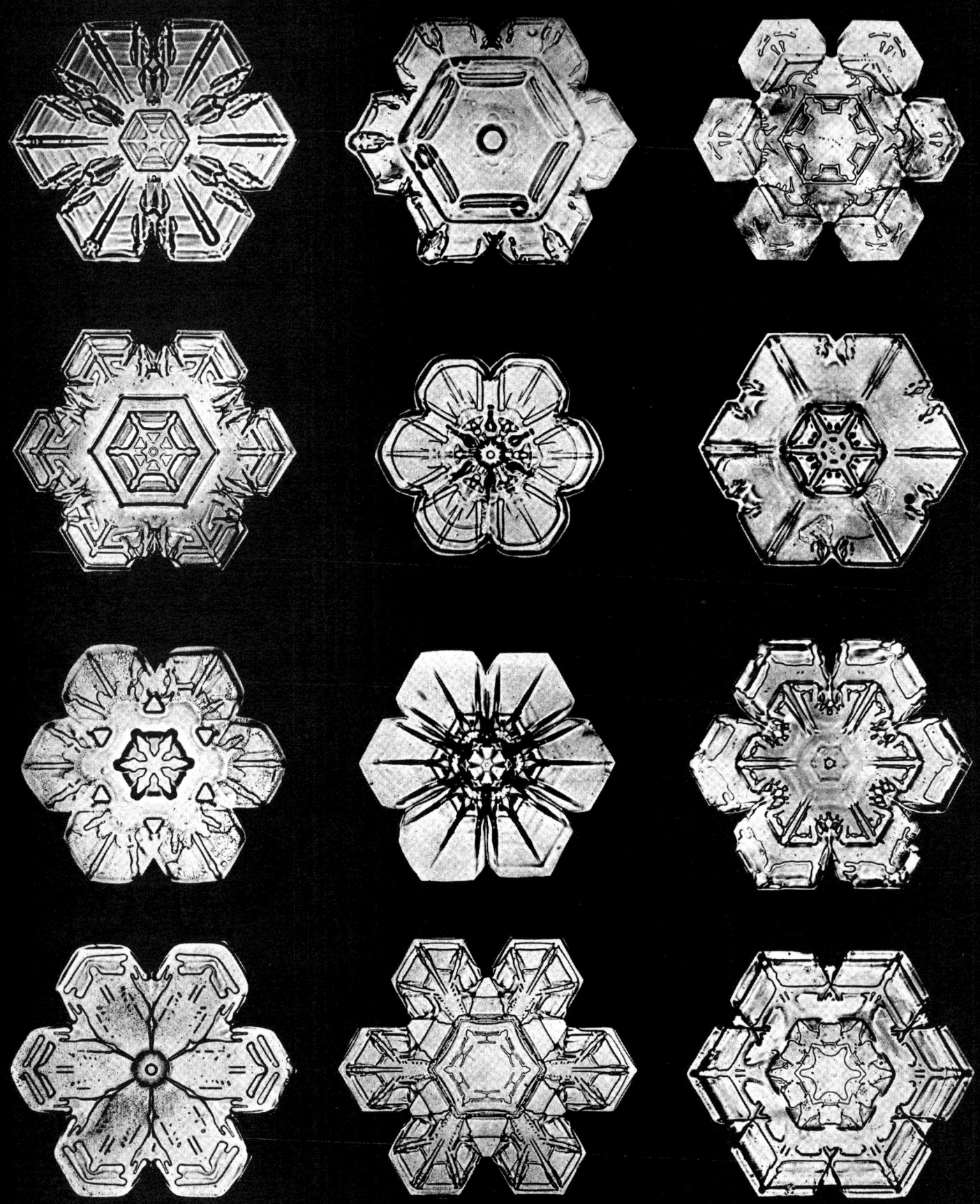

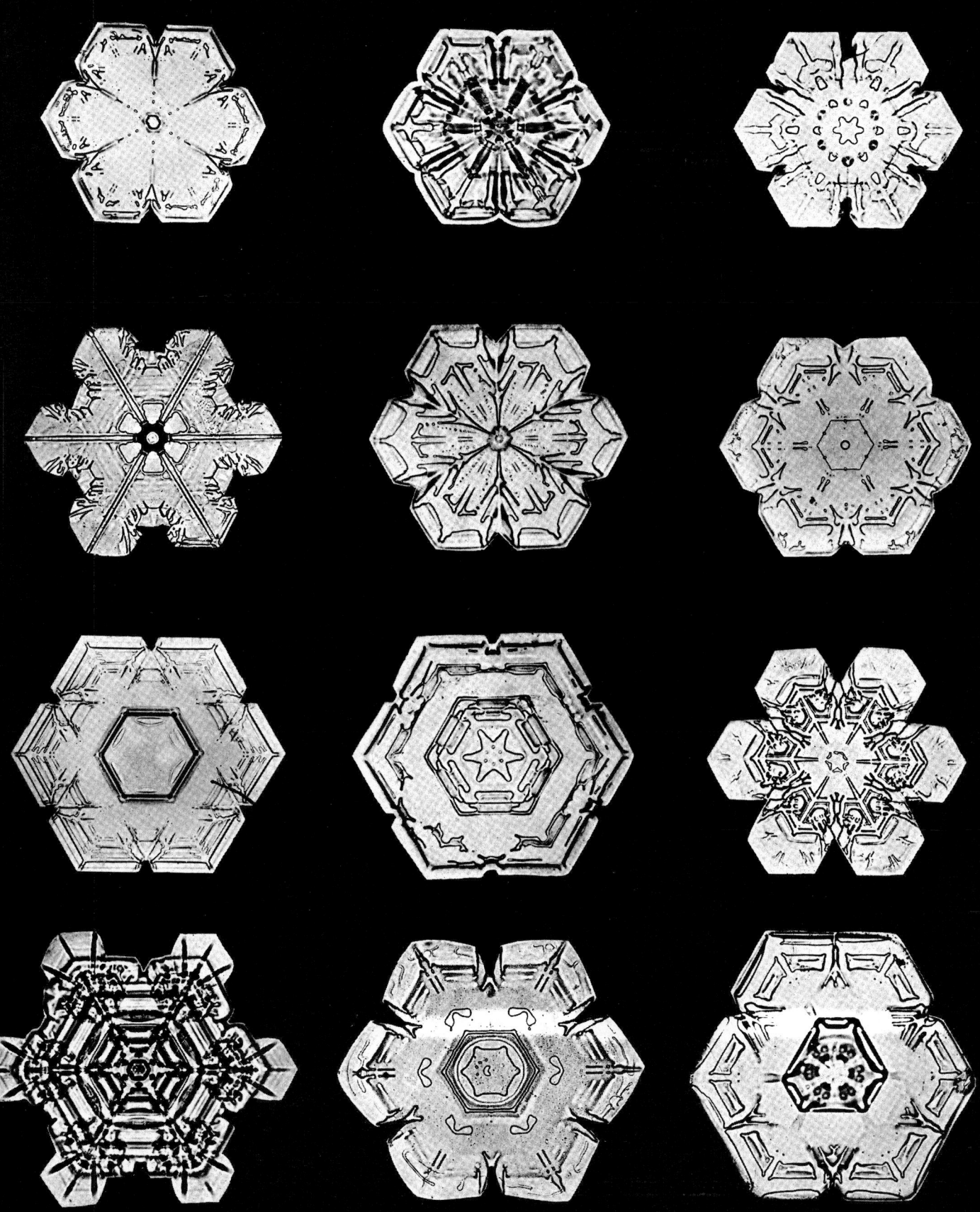

9

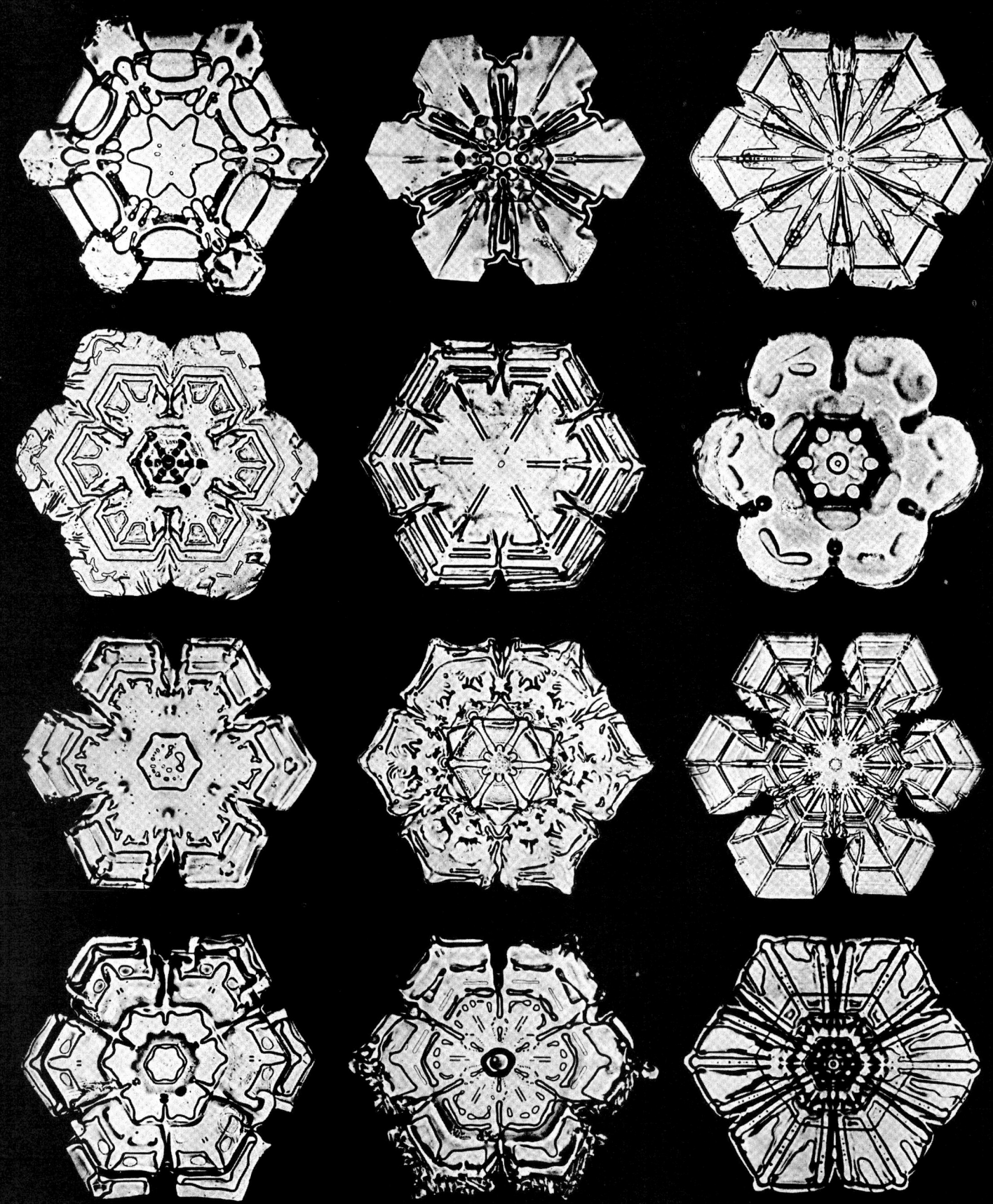

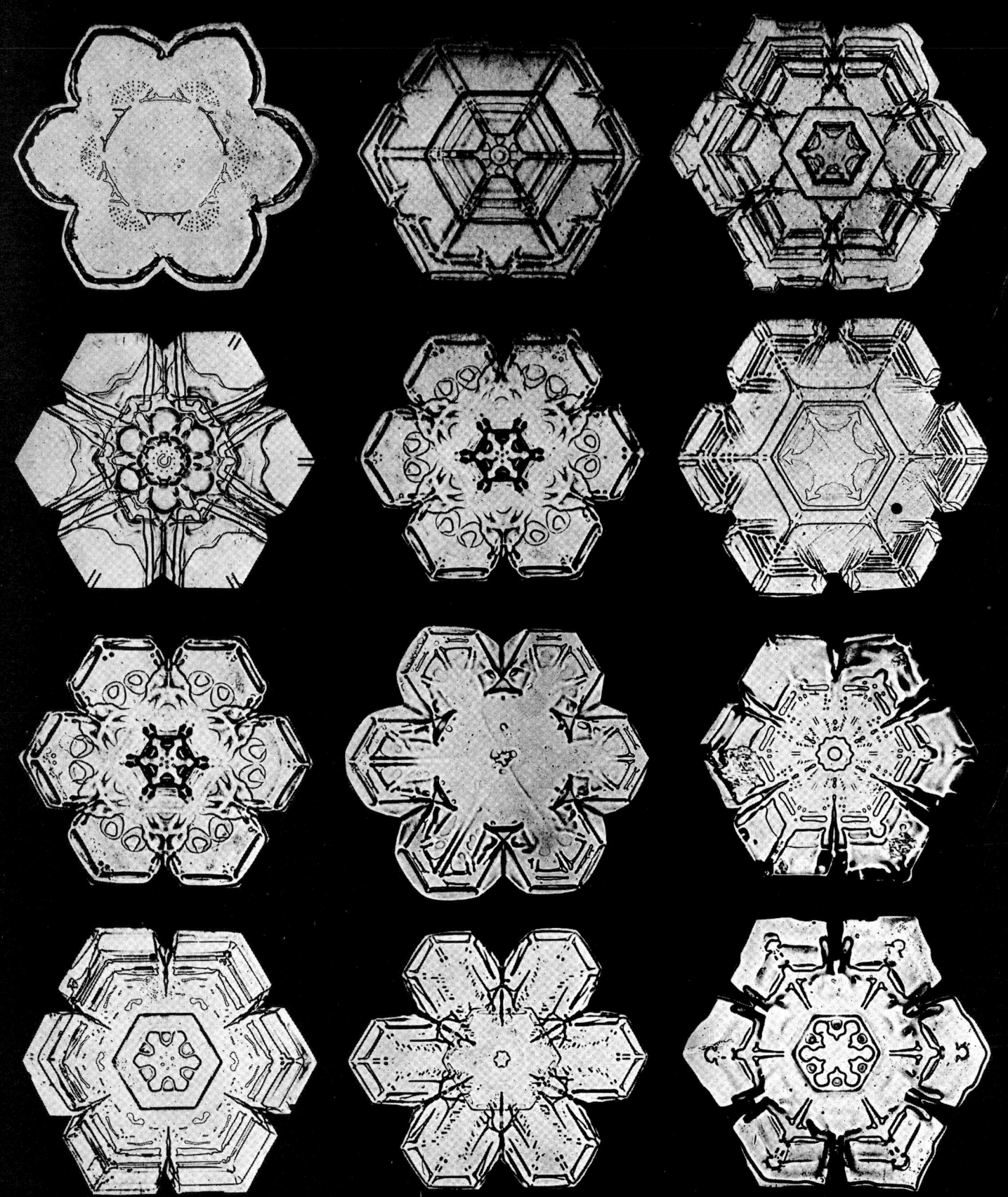

11

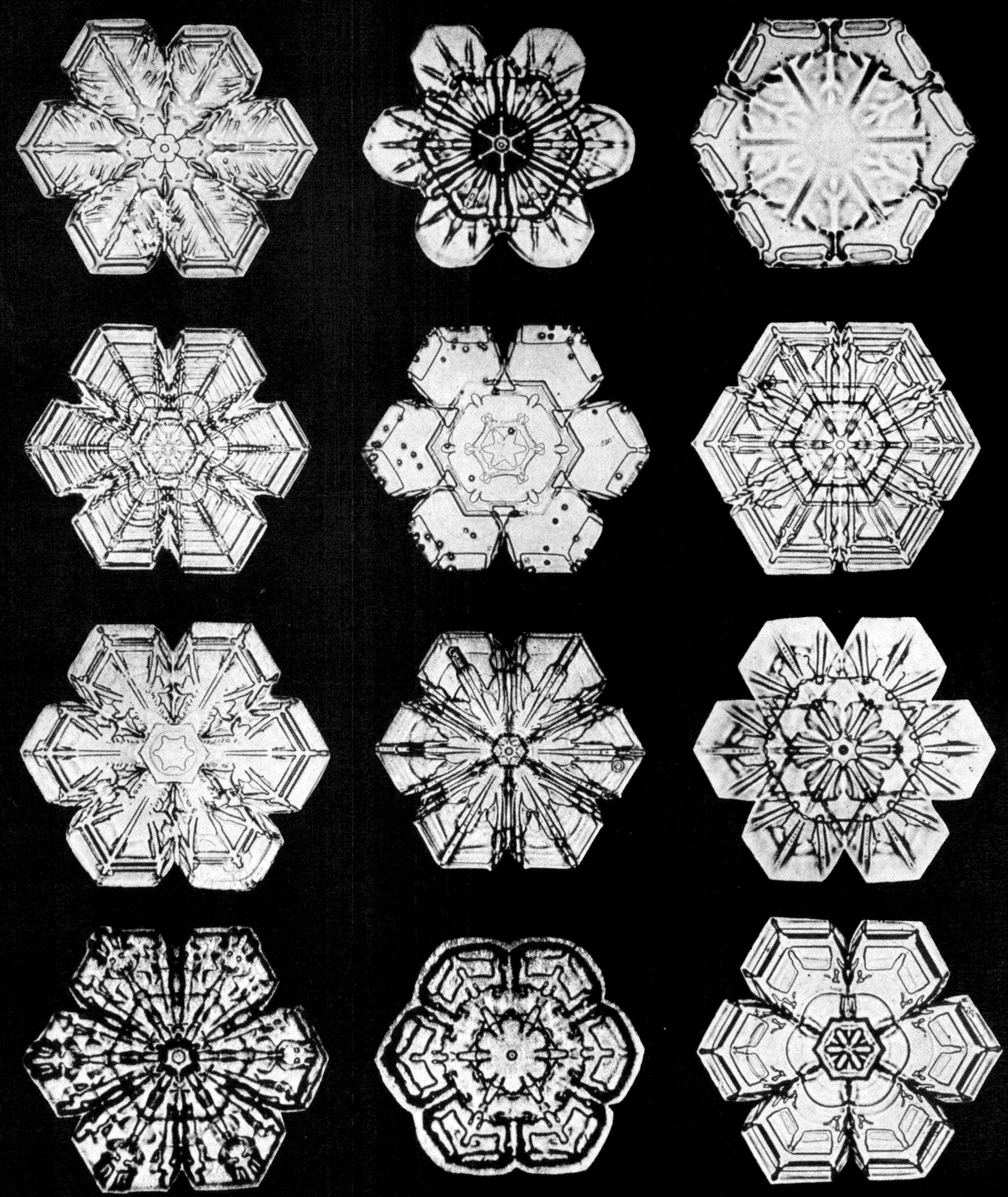

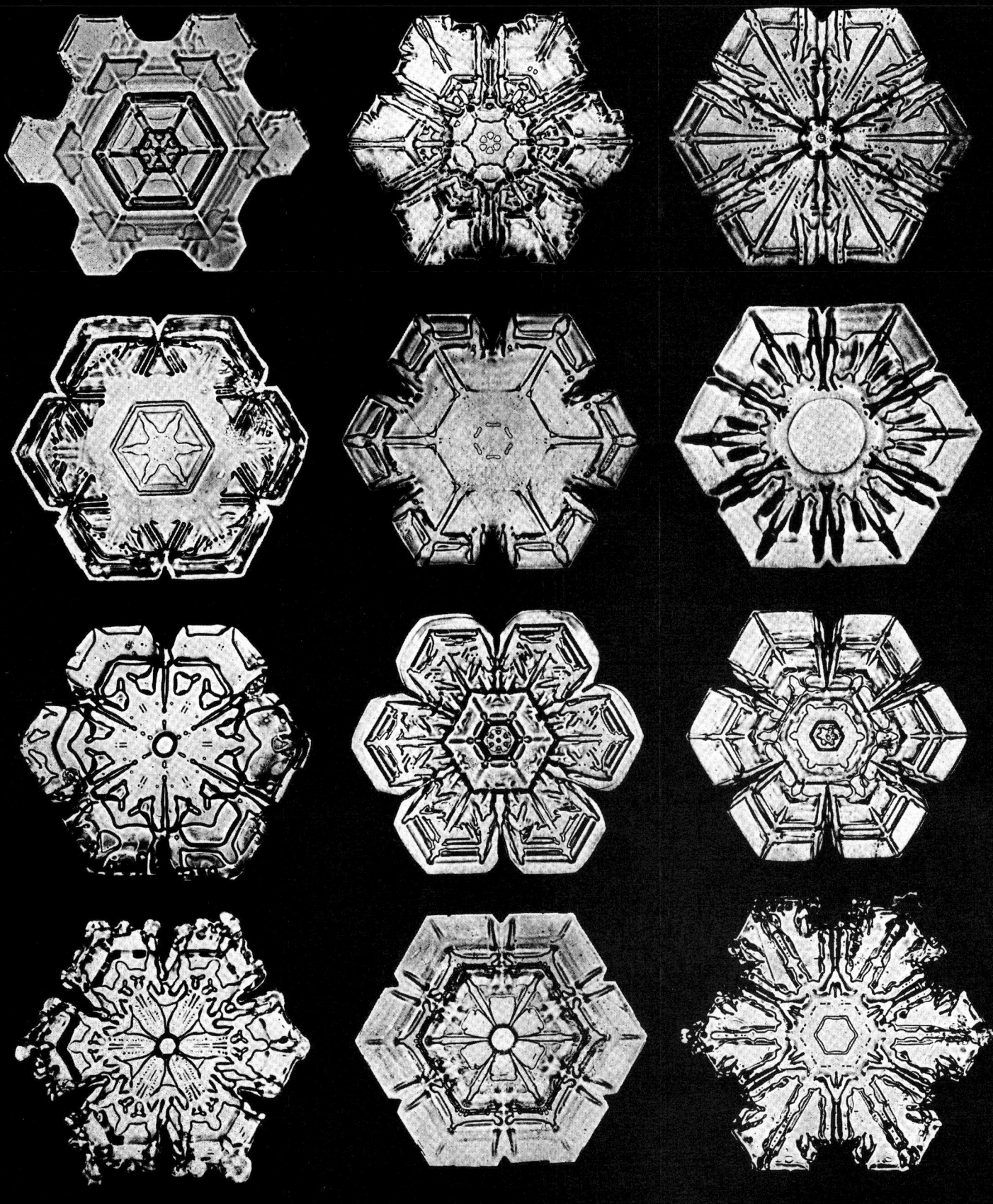

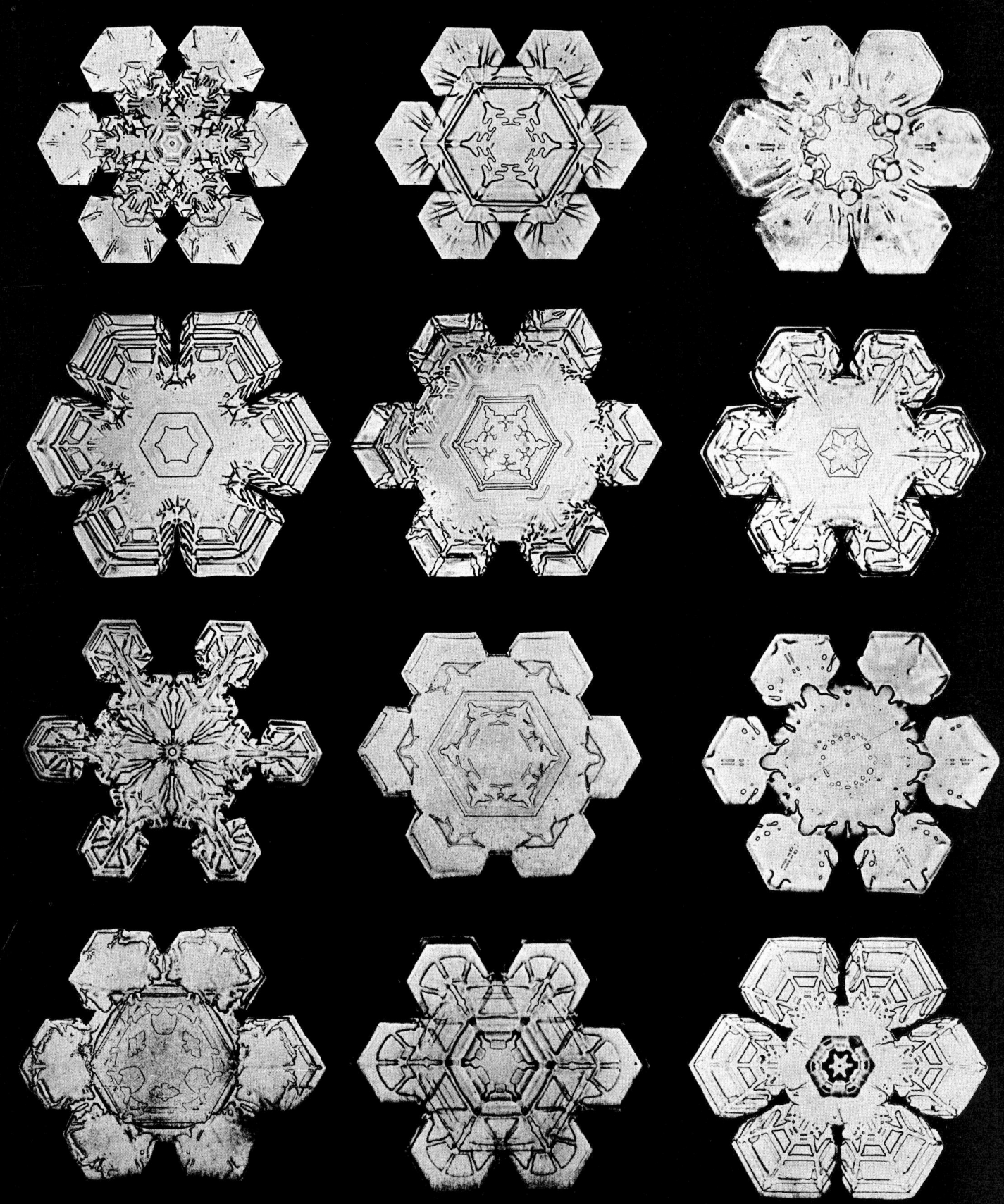

14

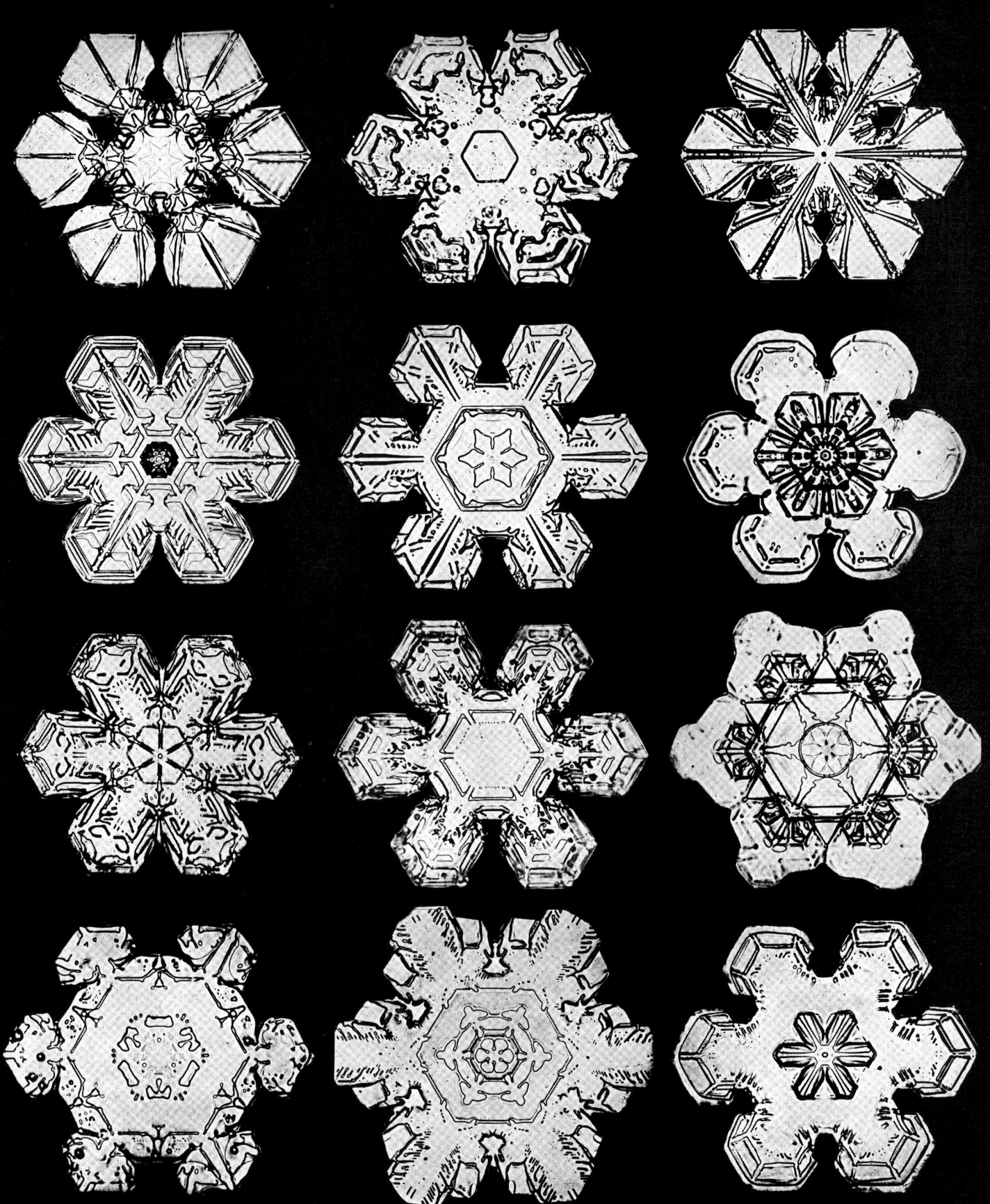

15

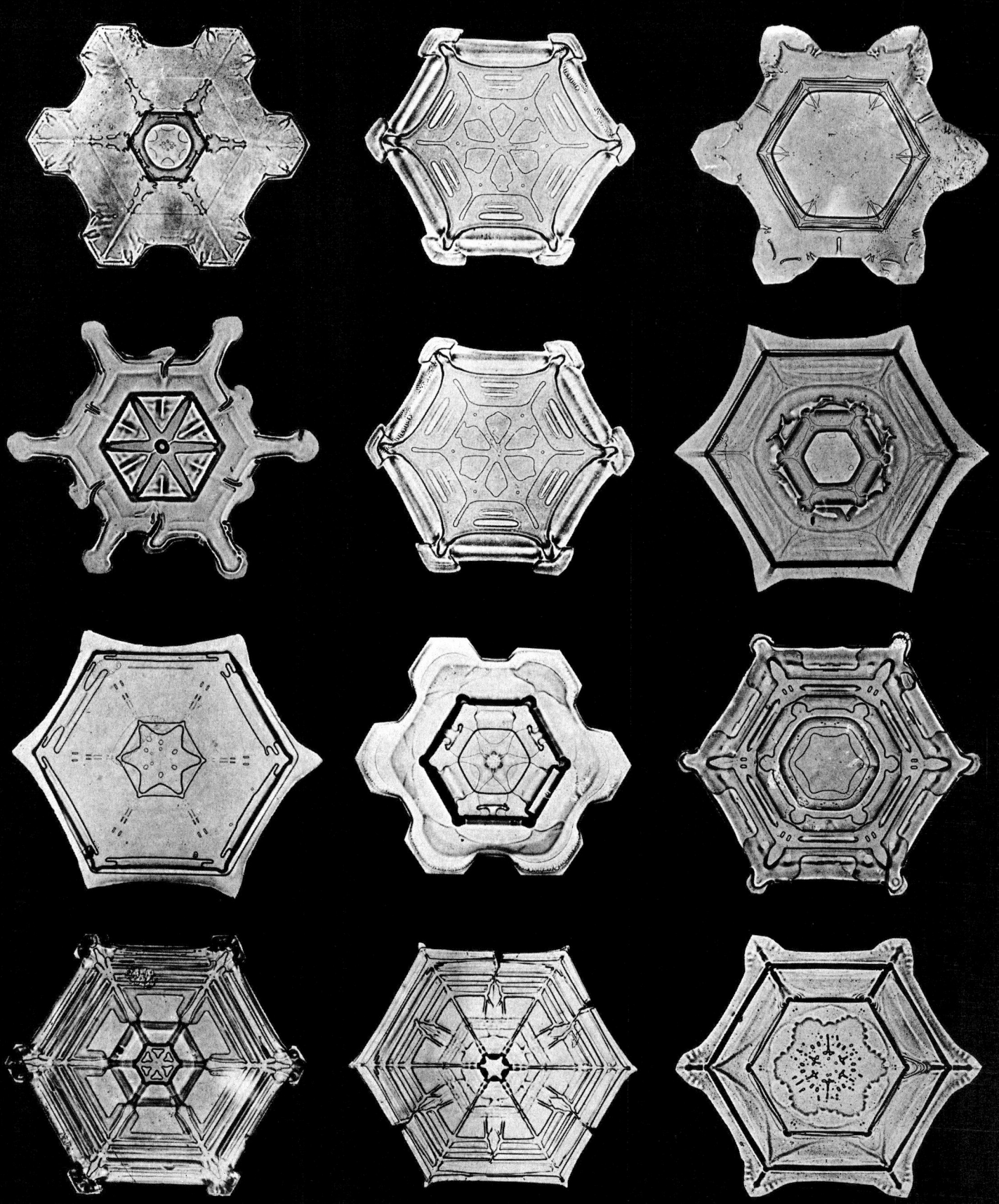

16

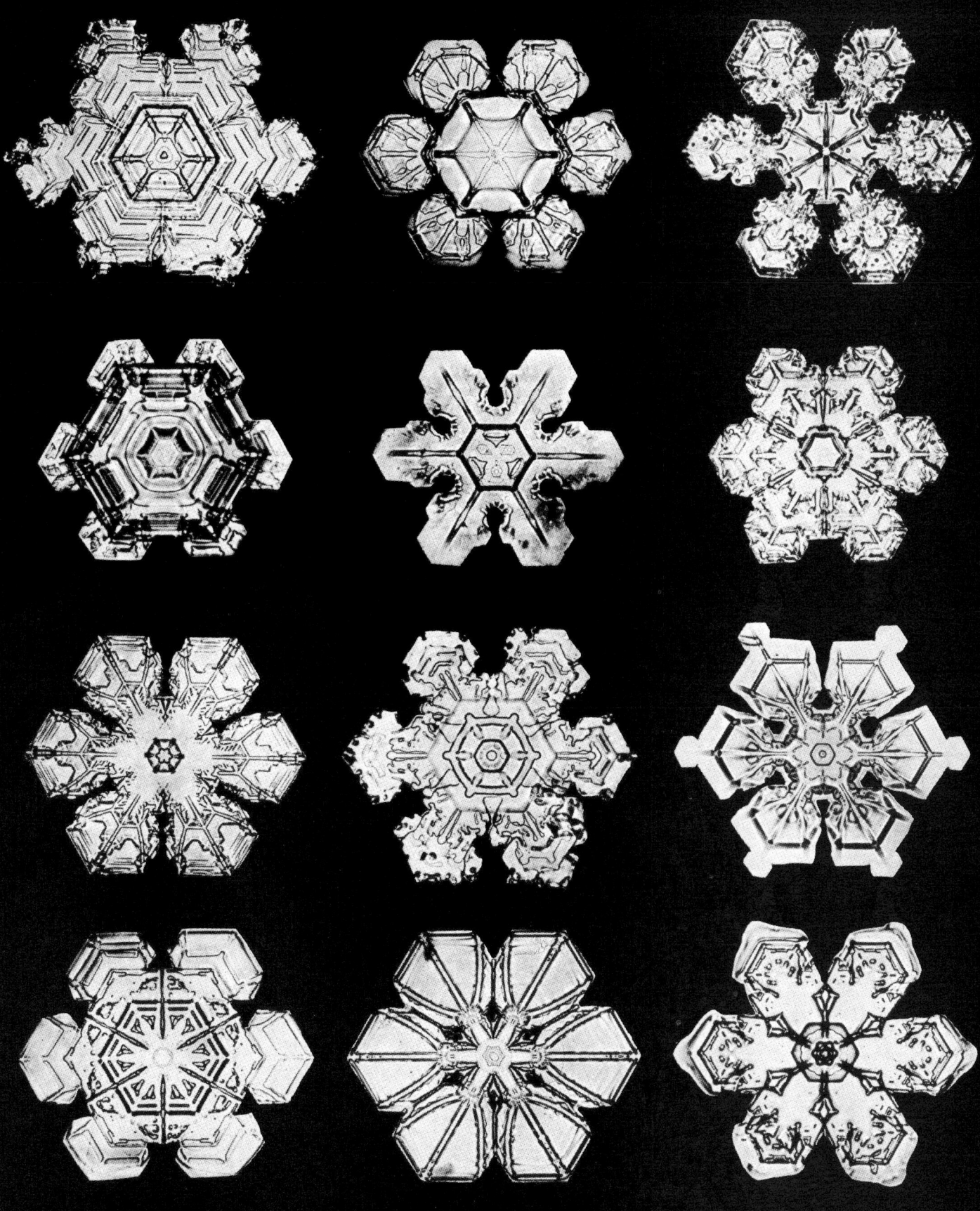

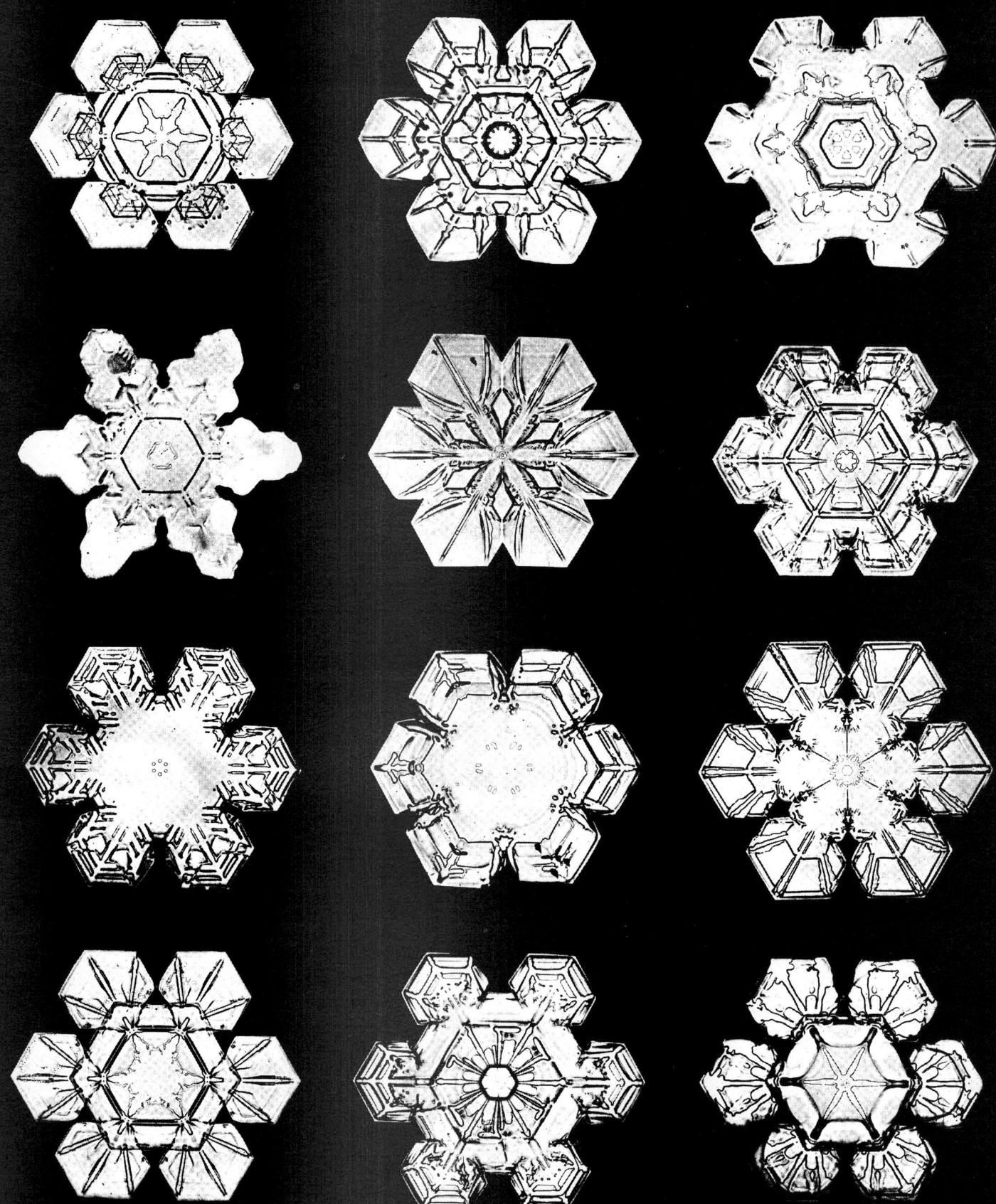

18

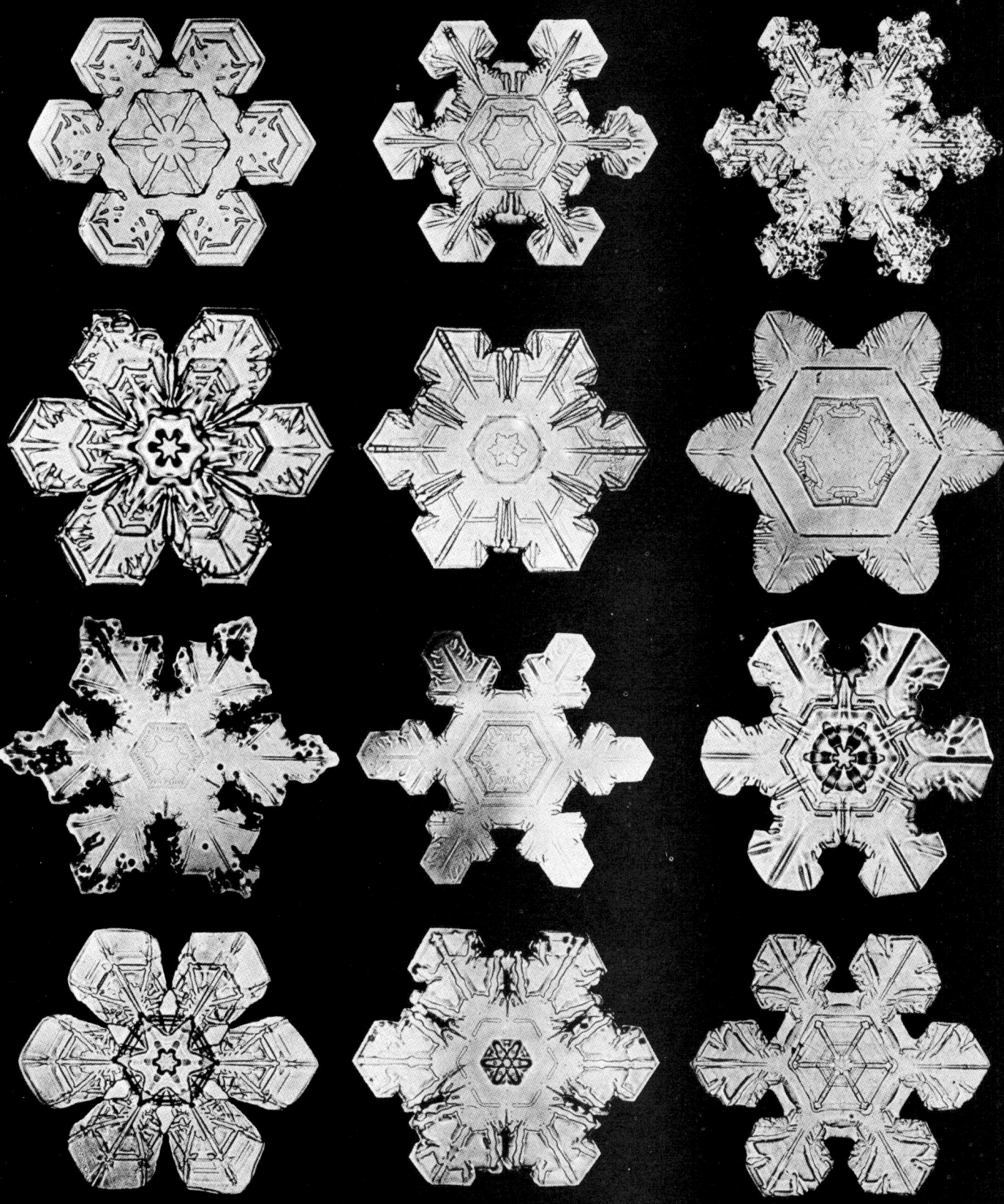

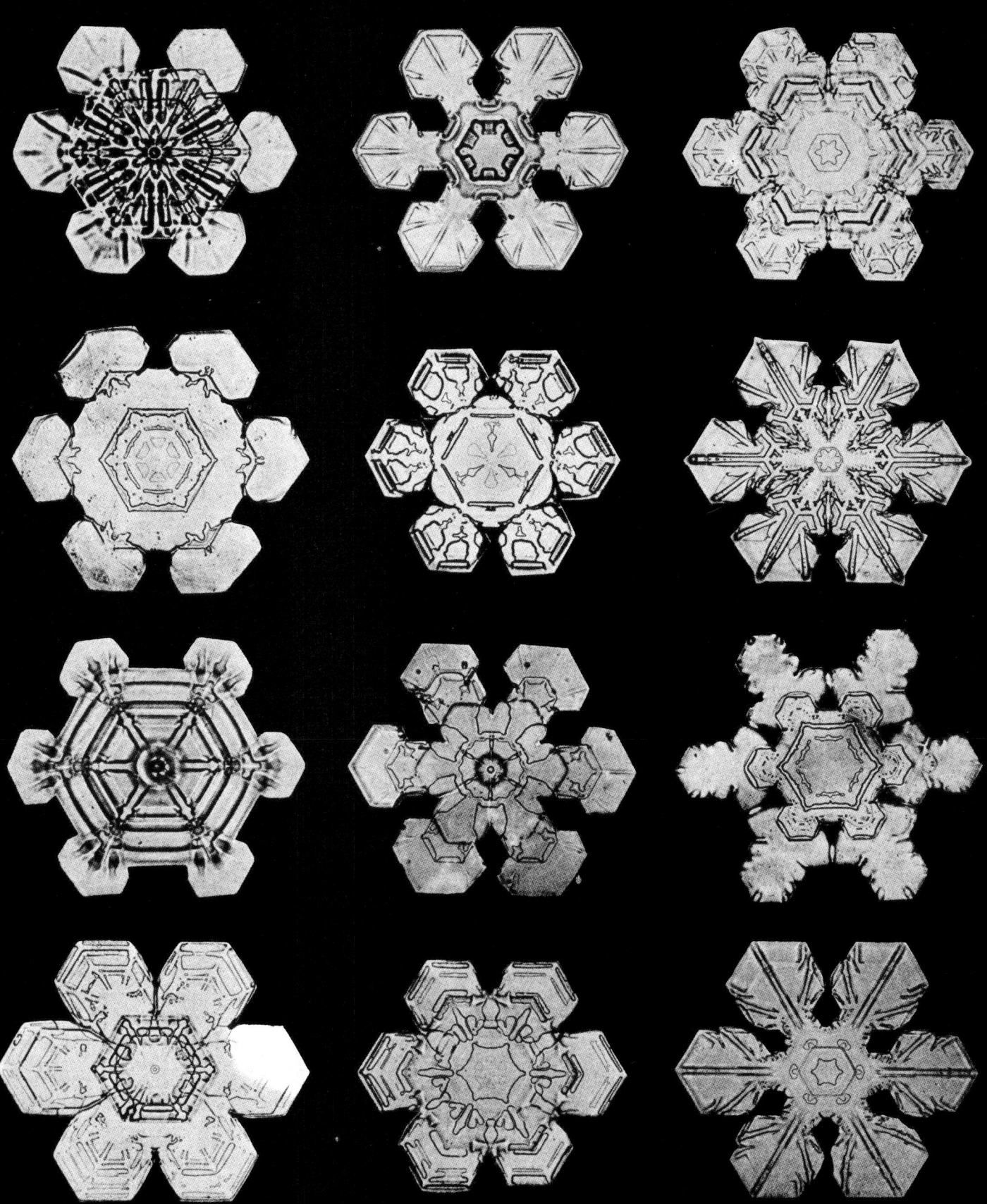

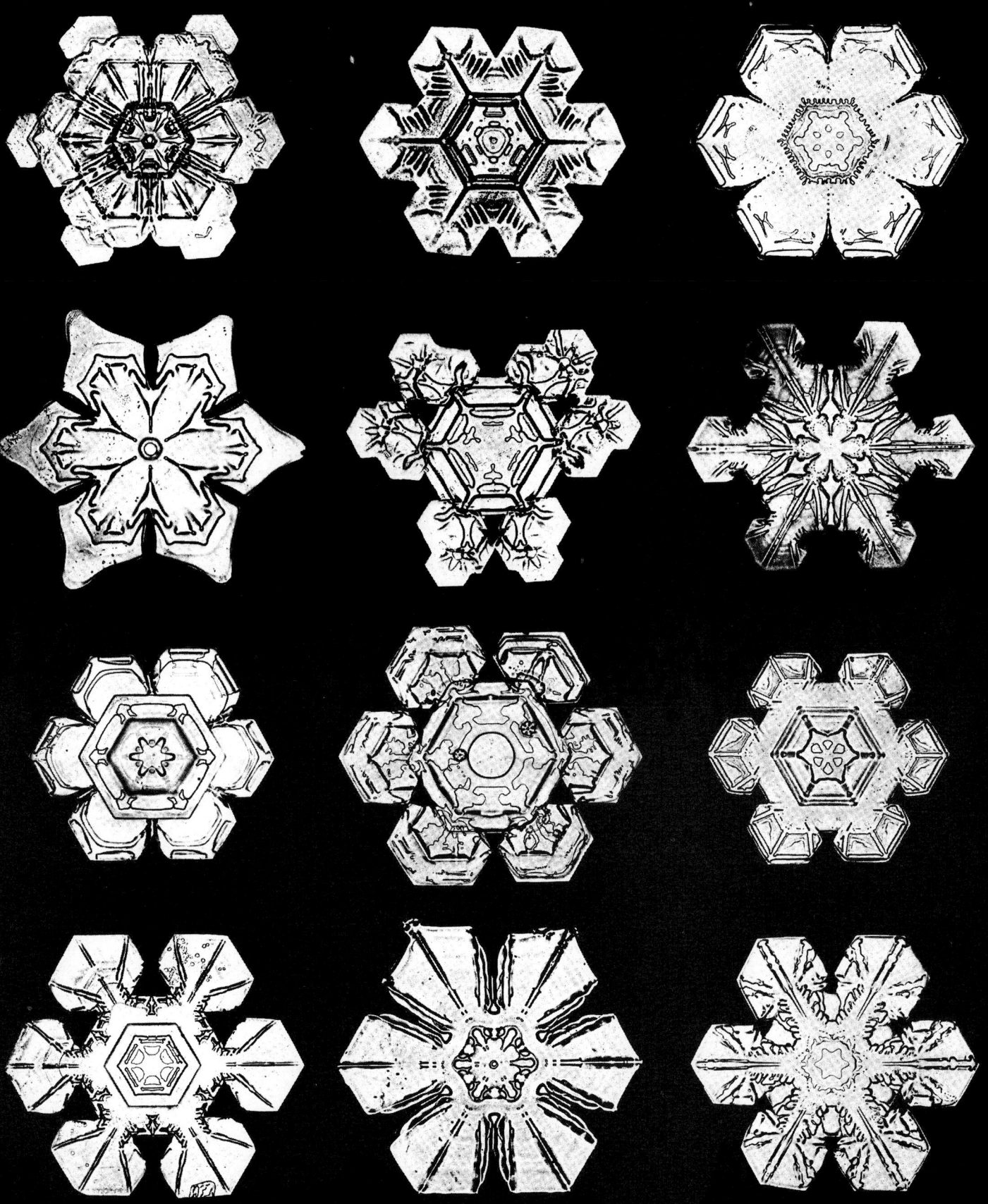

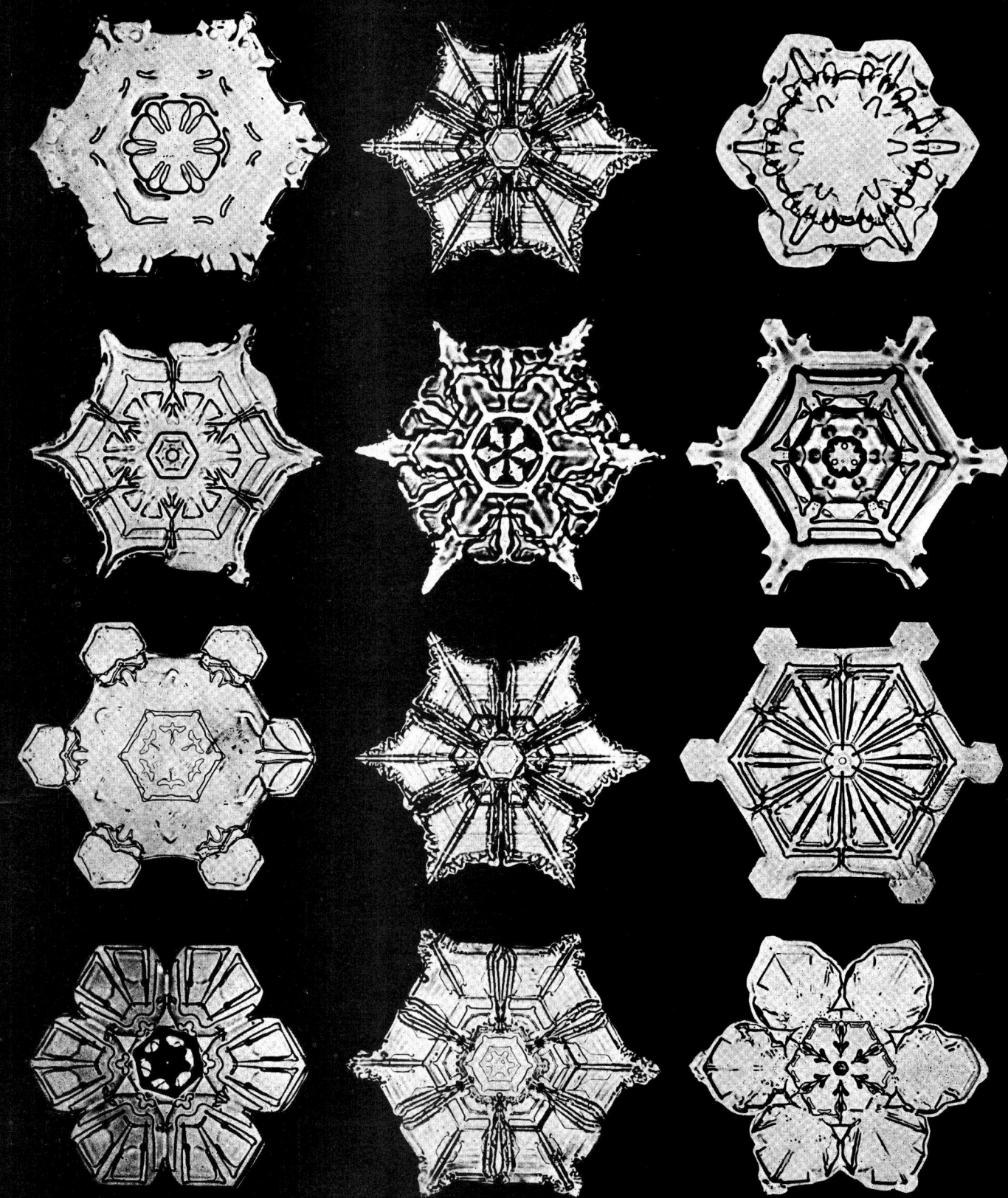

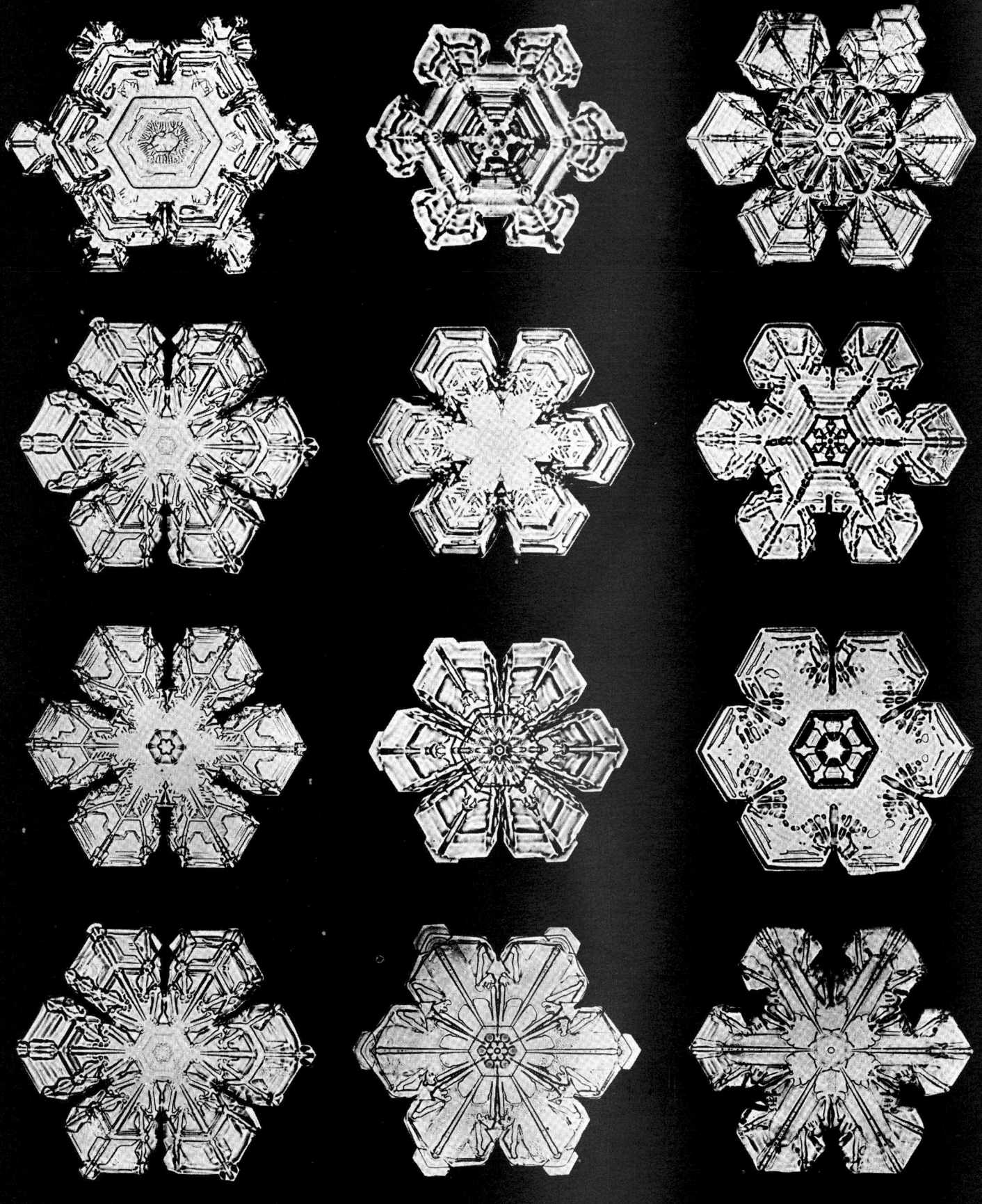

23

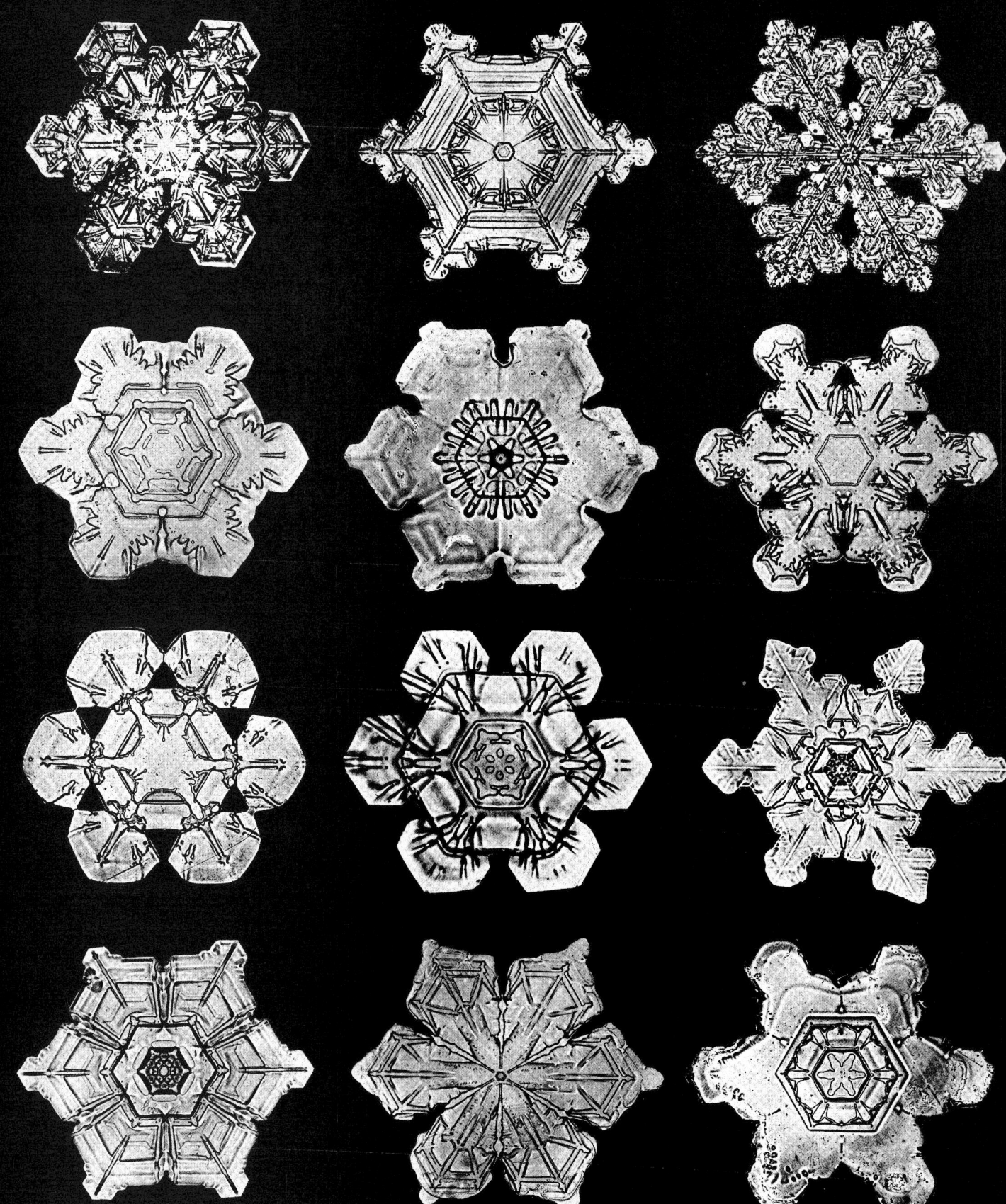

24

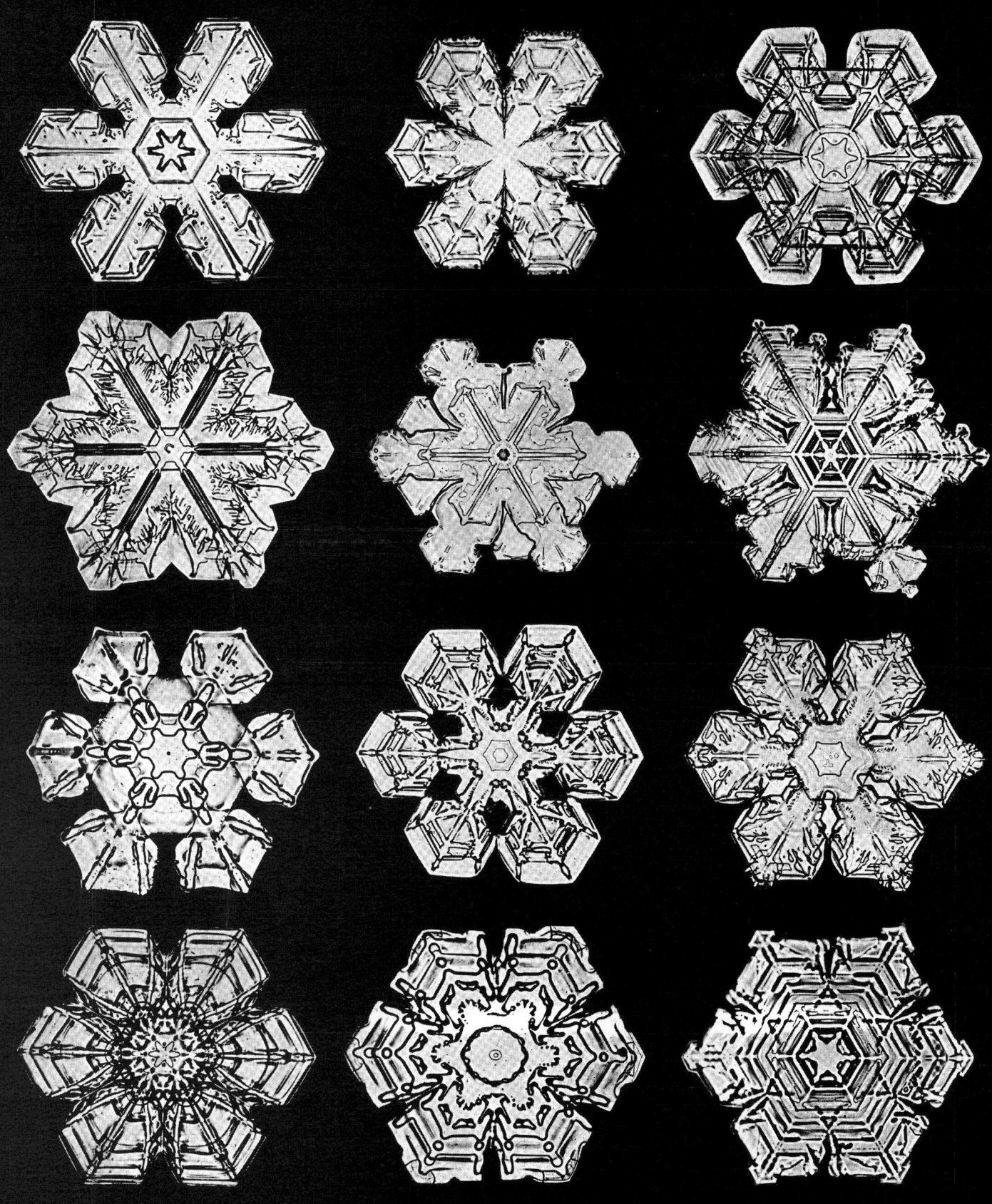

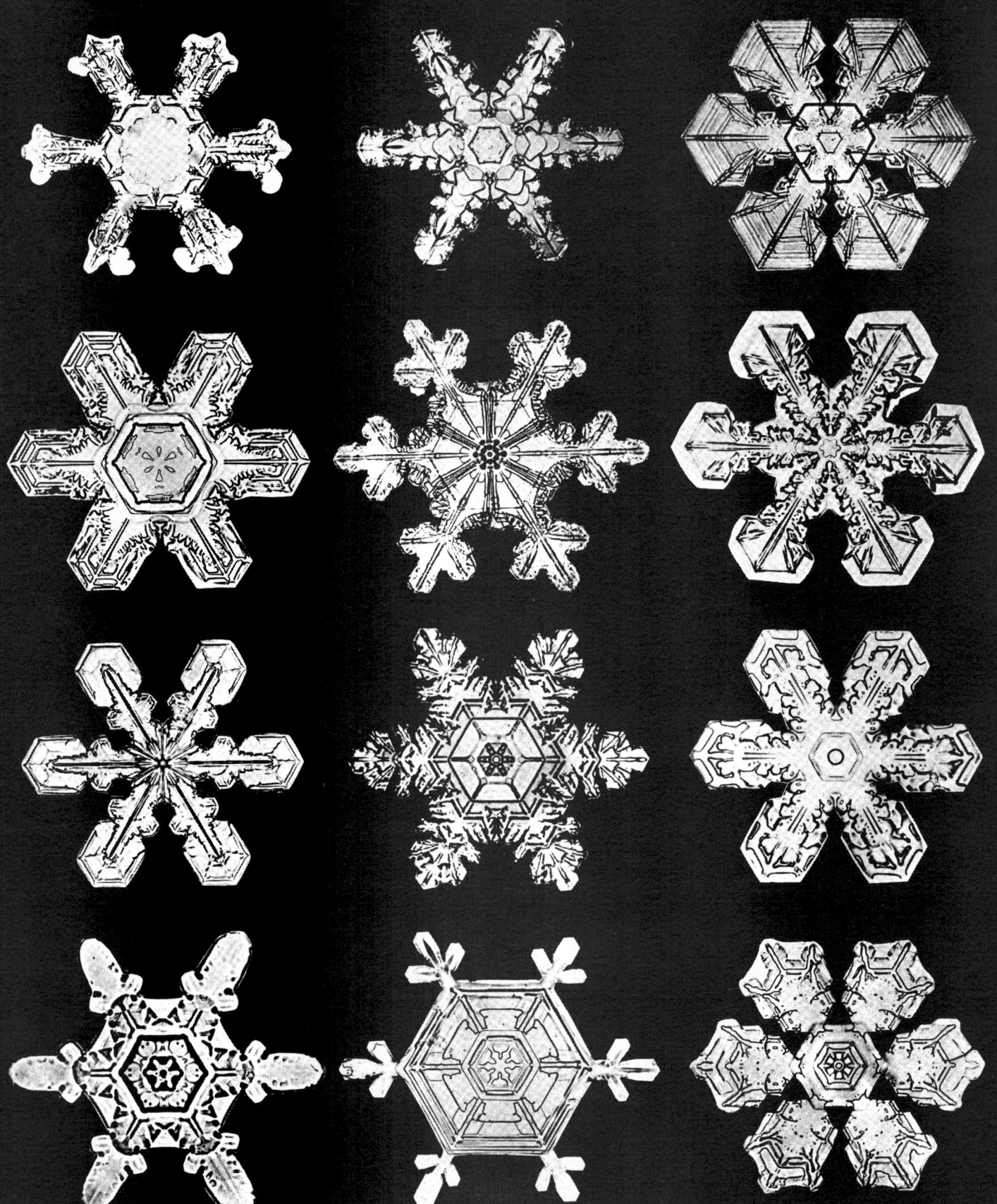

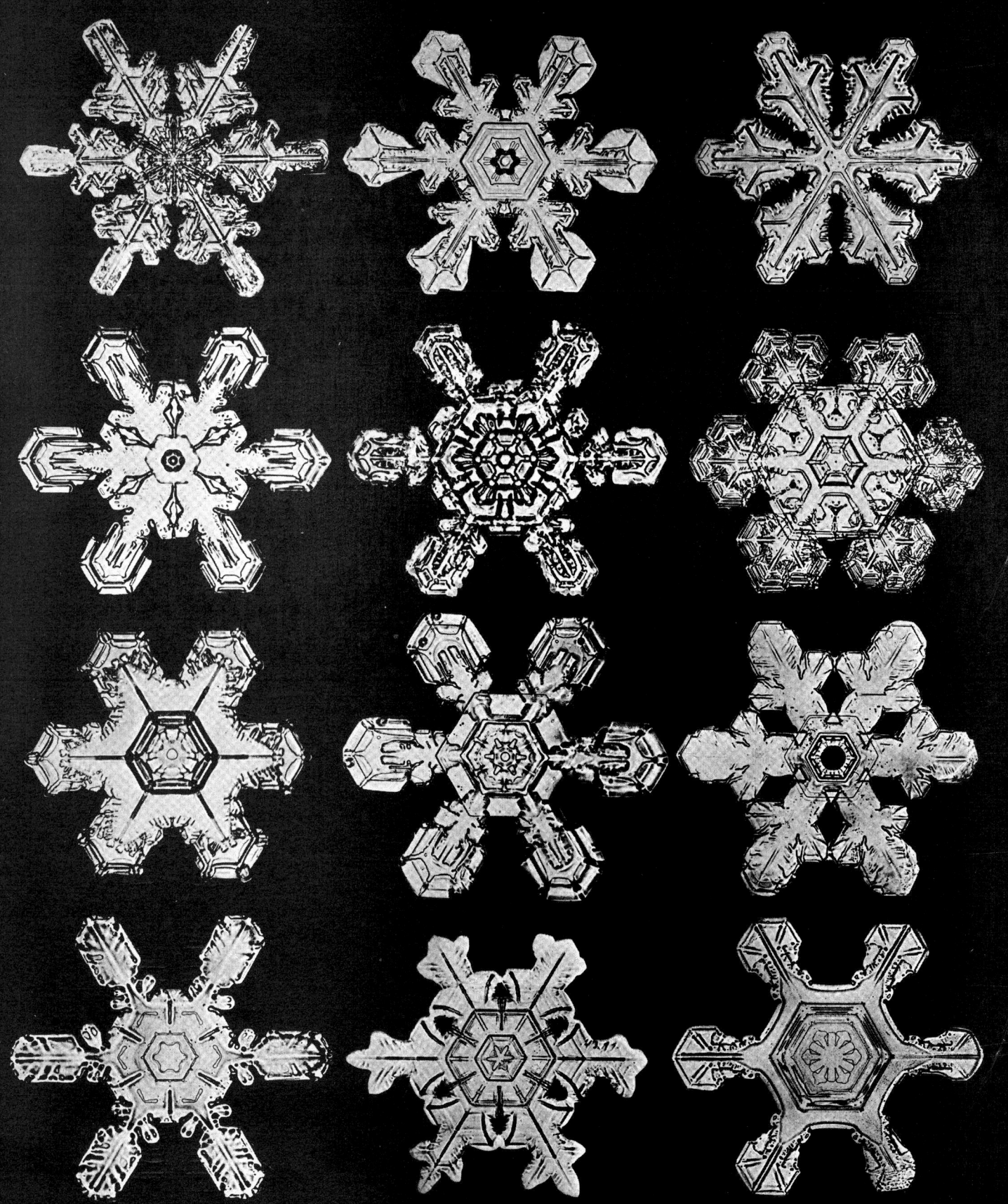

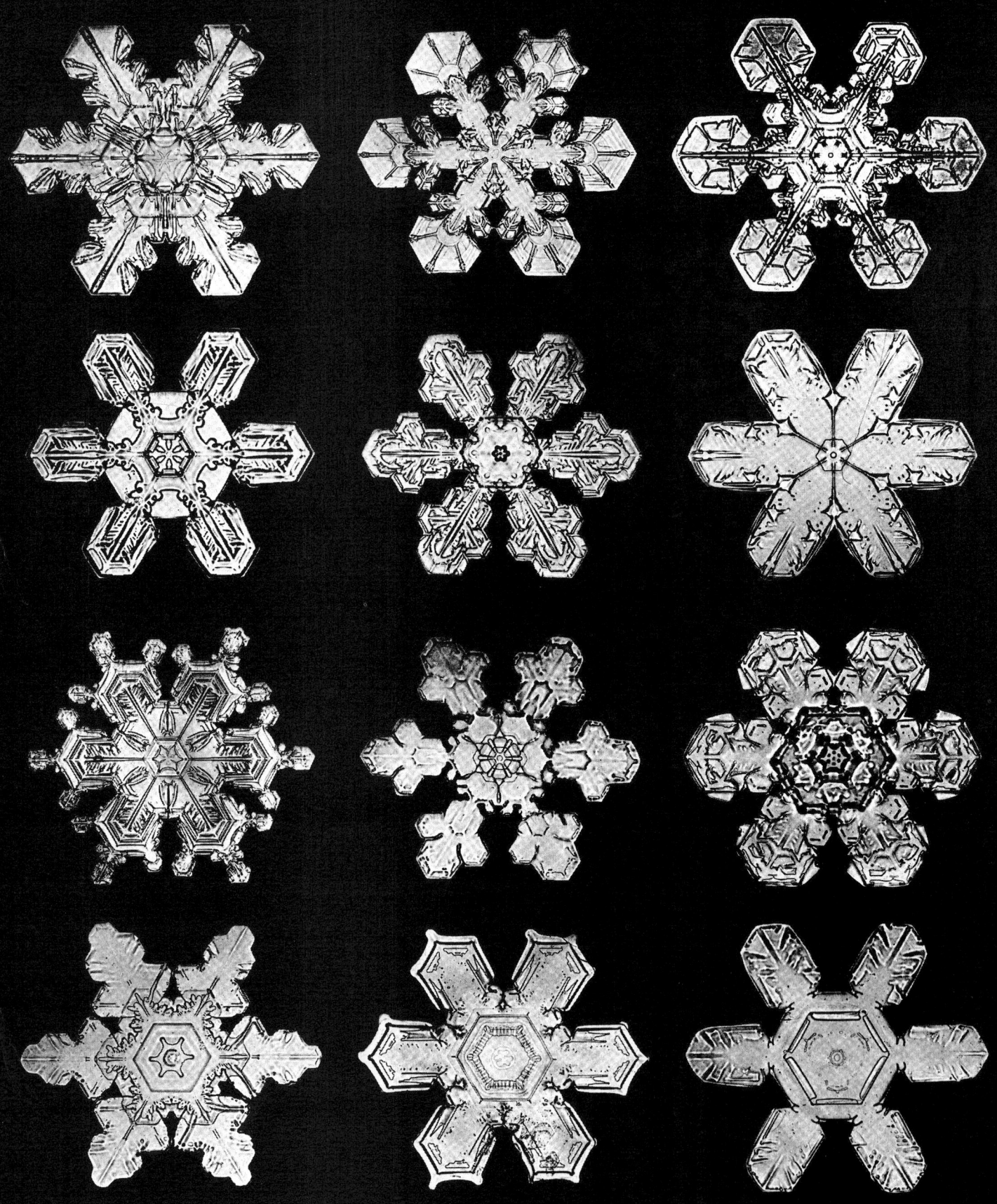

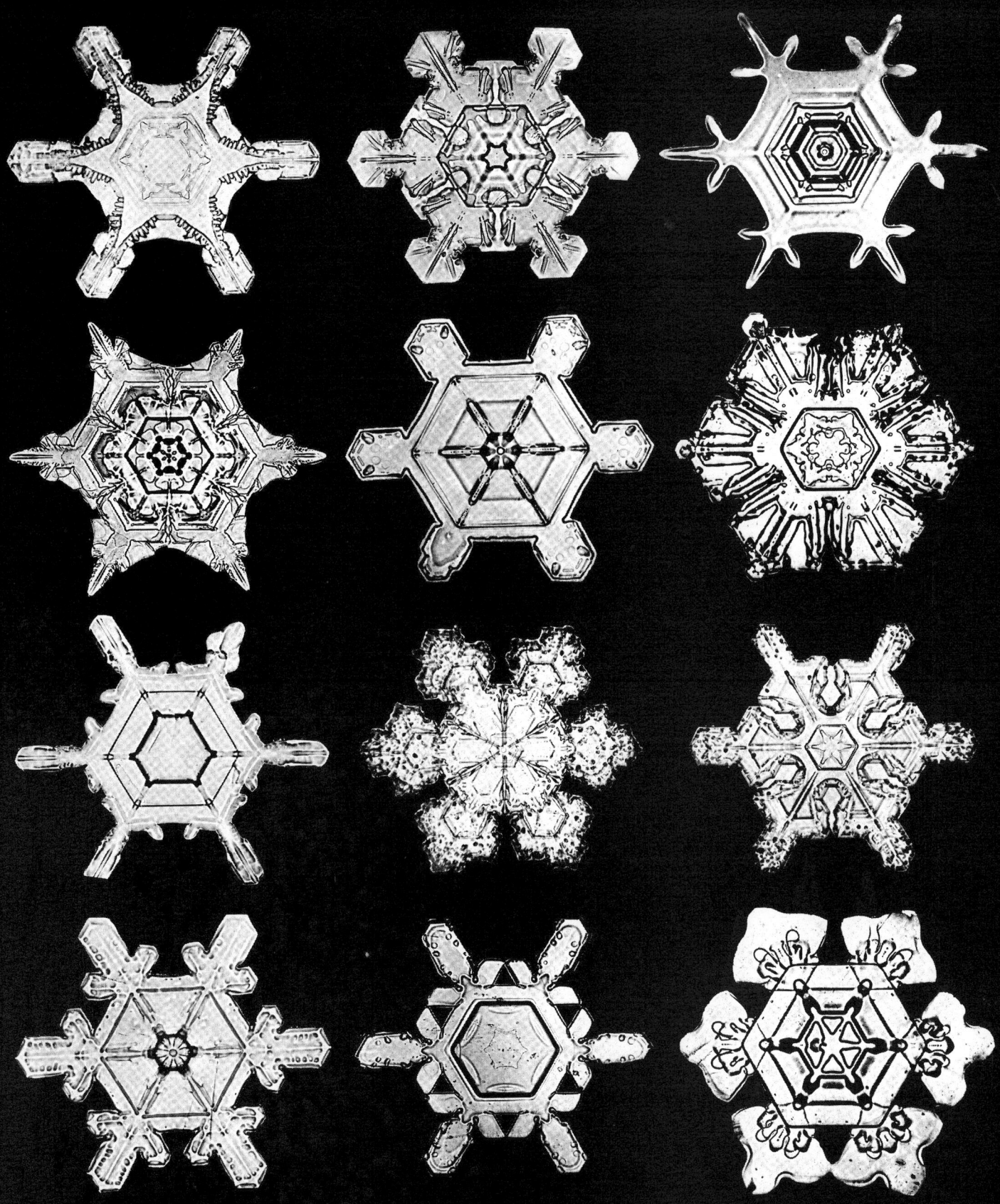

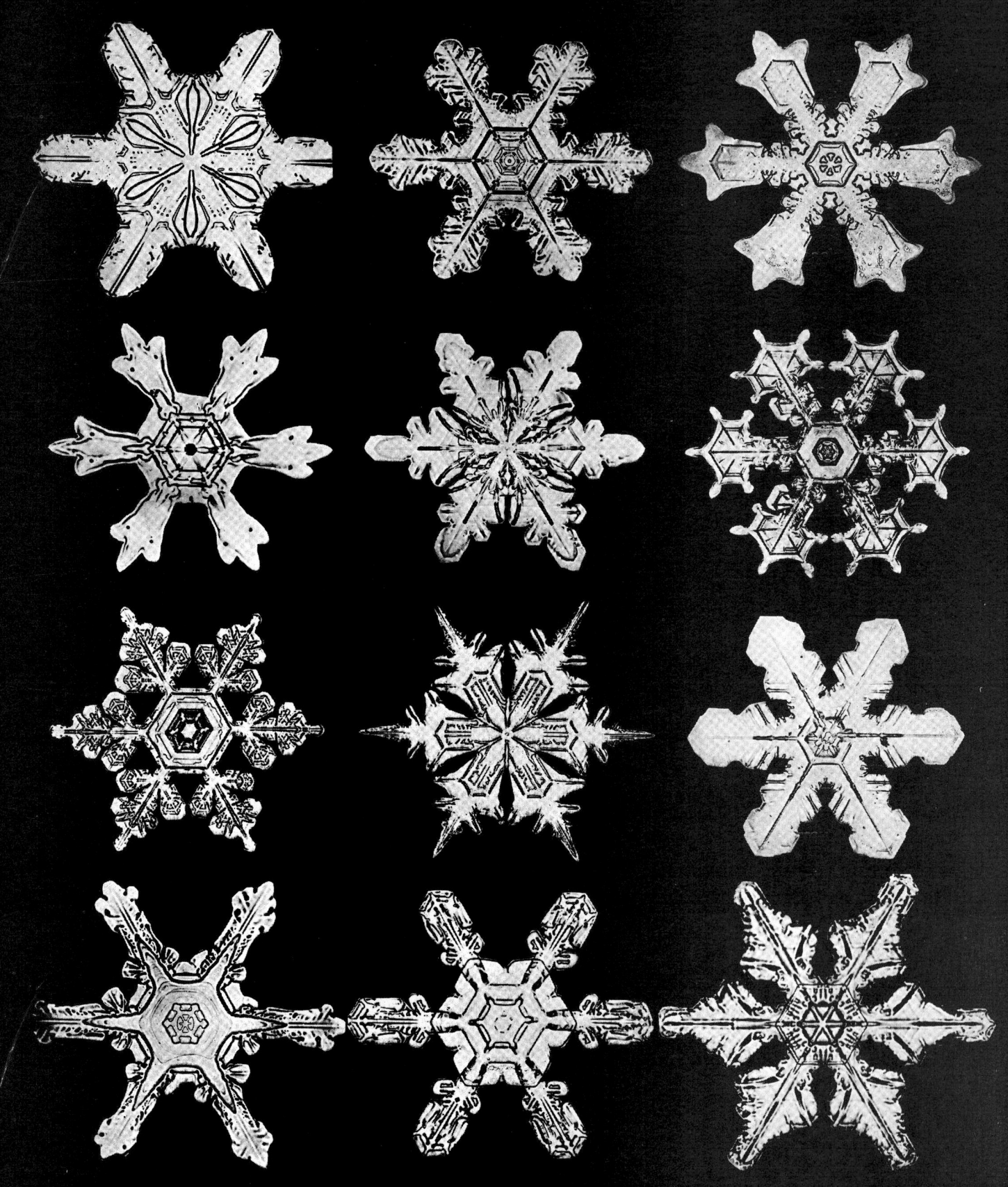

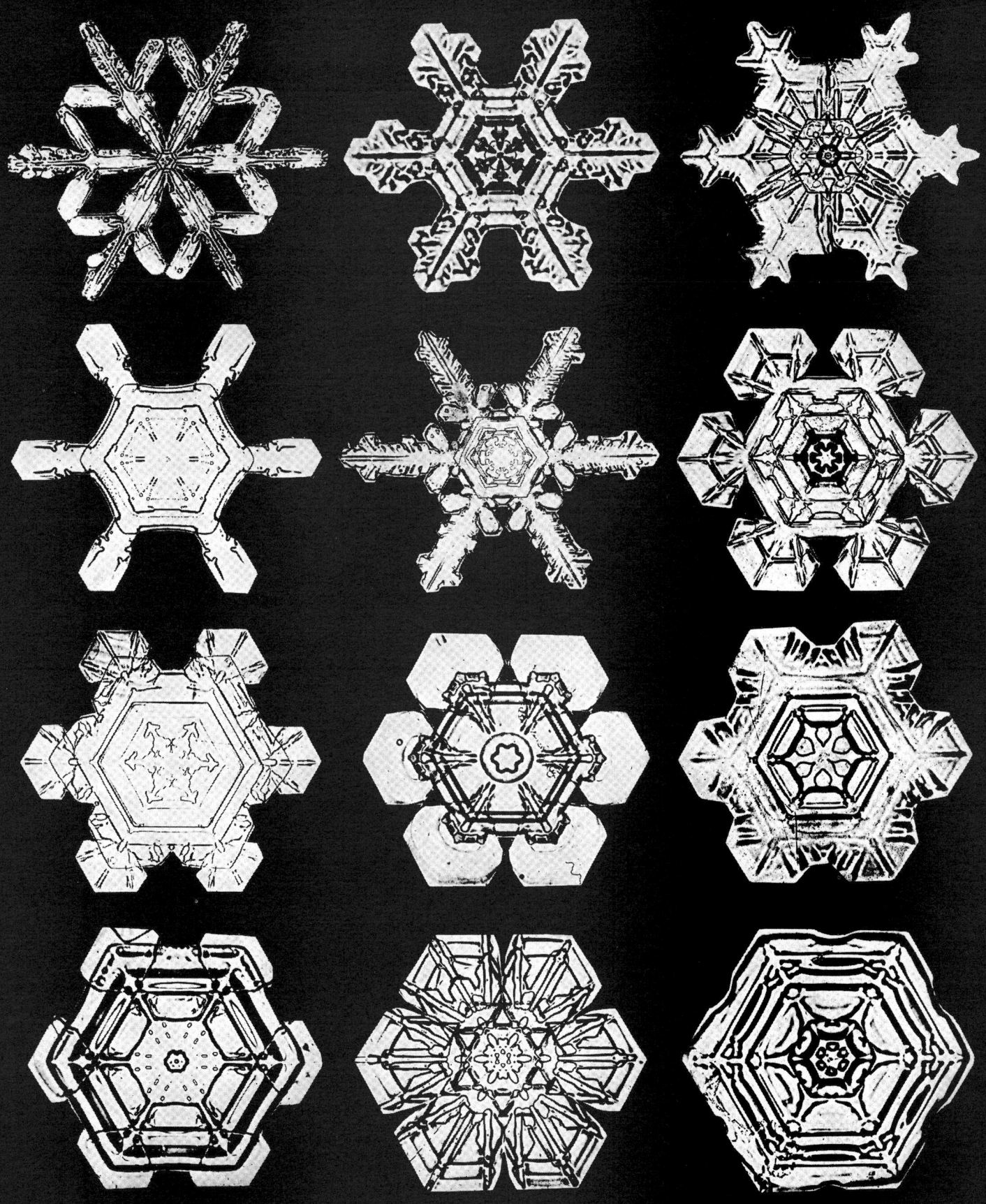

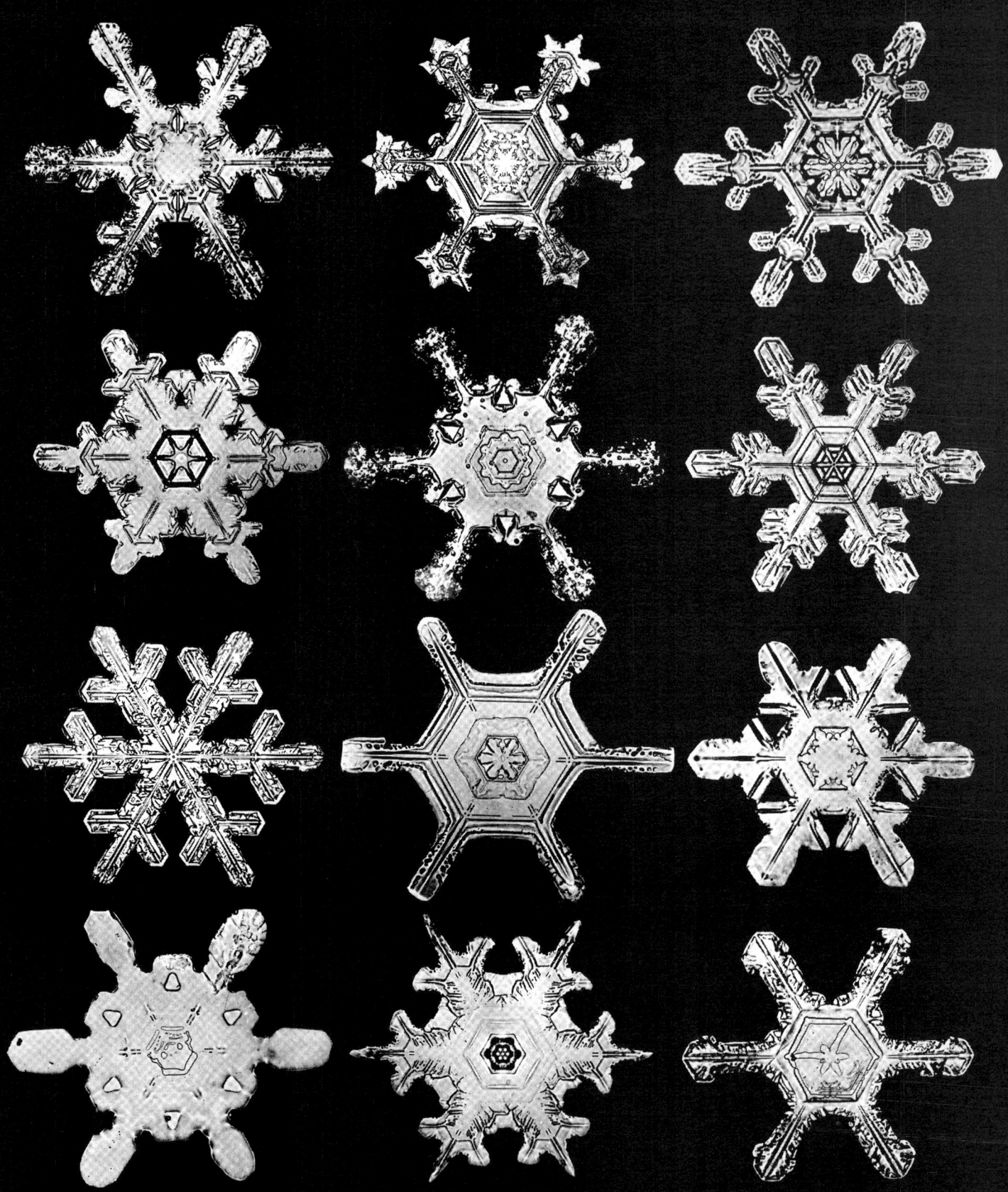

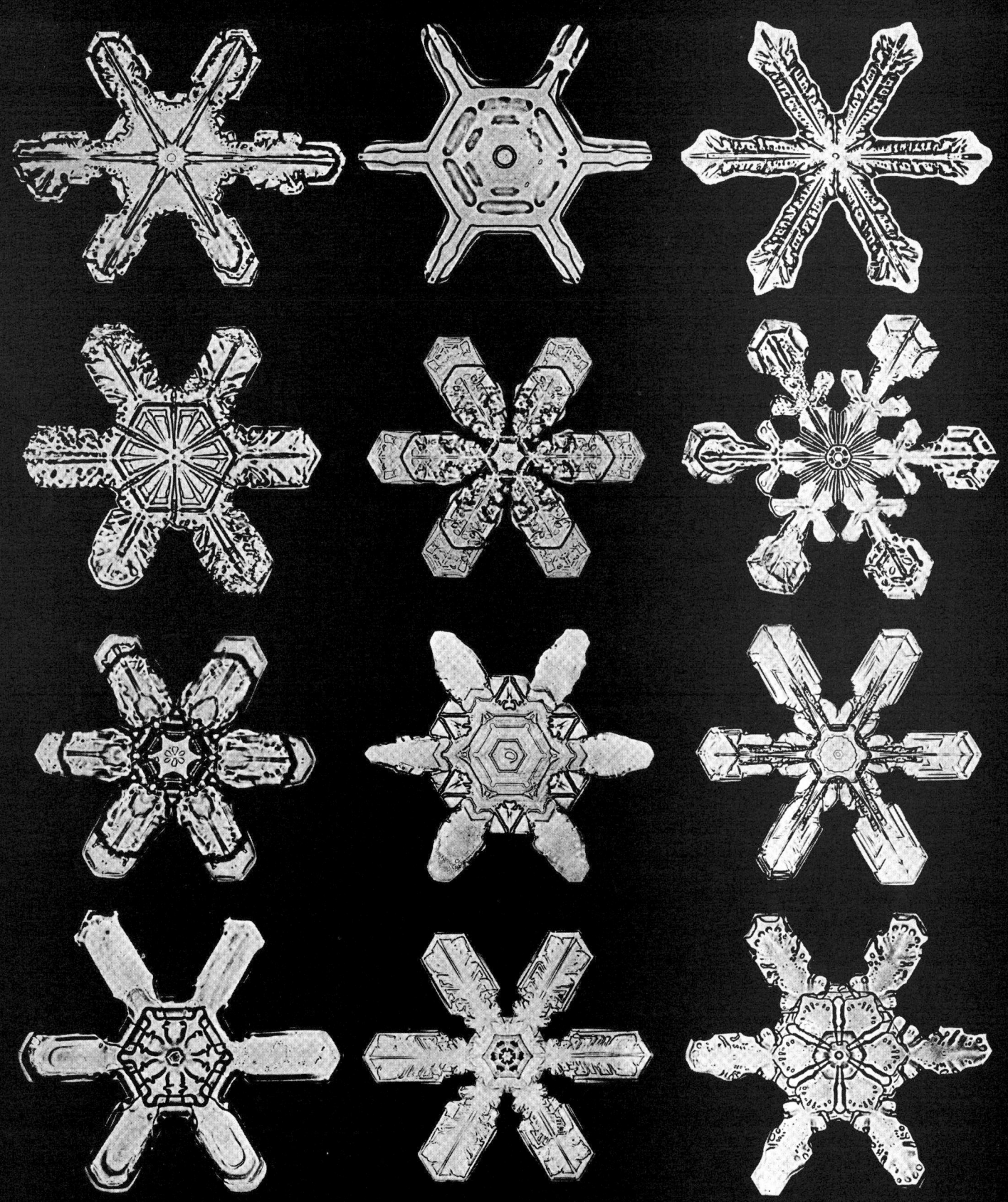

33

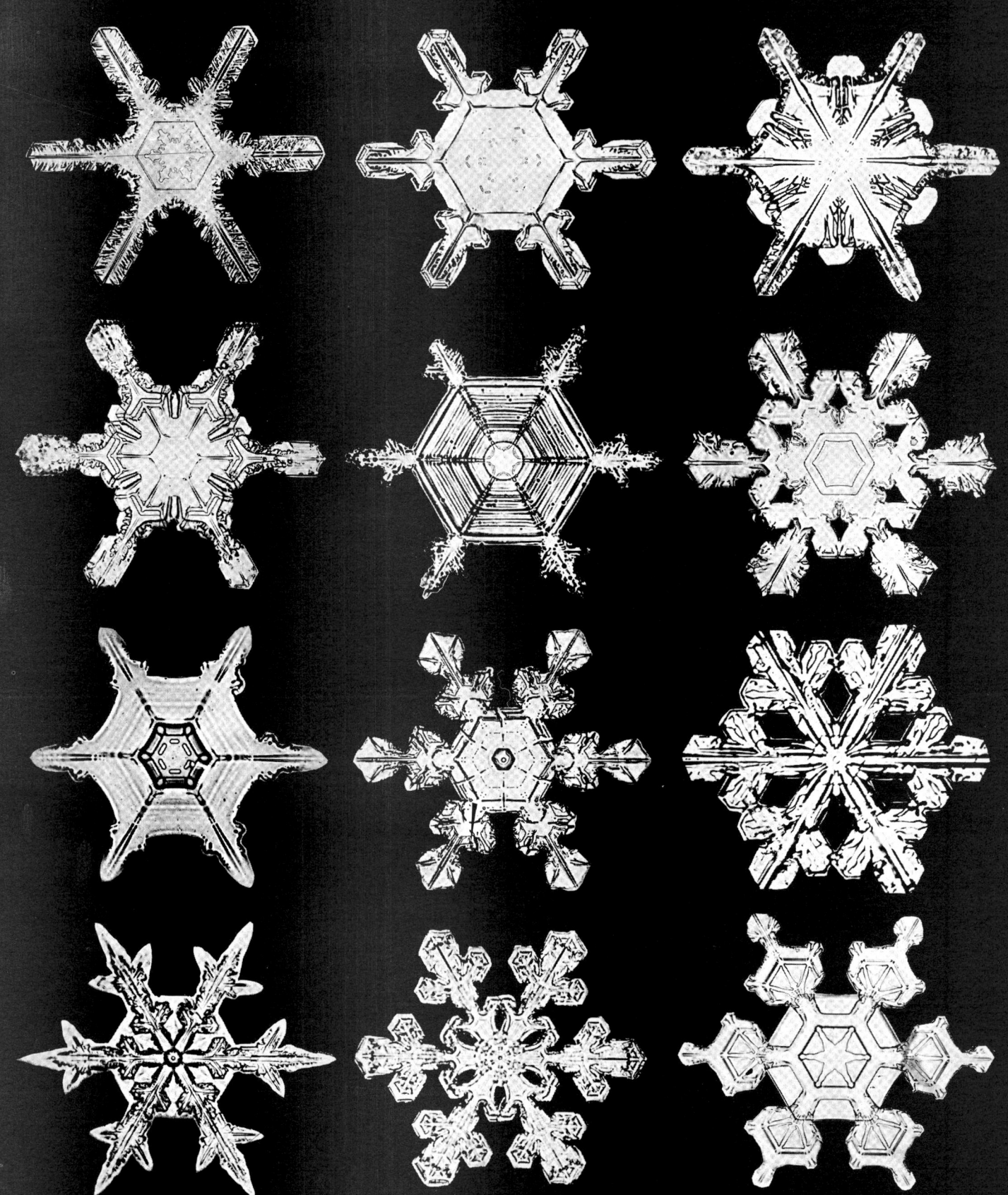

34

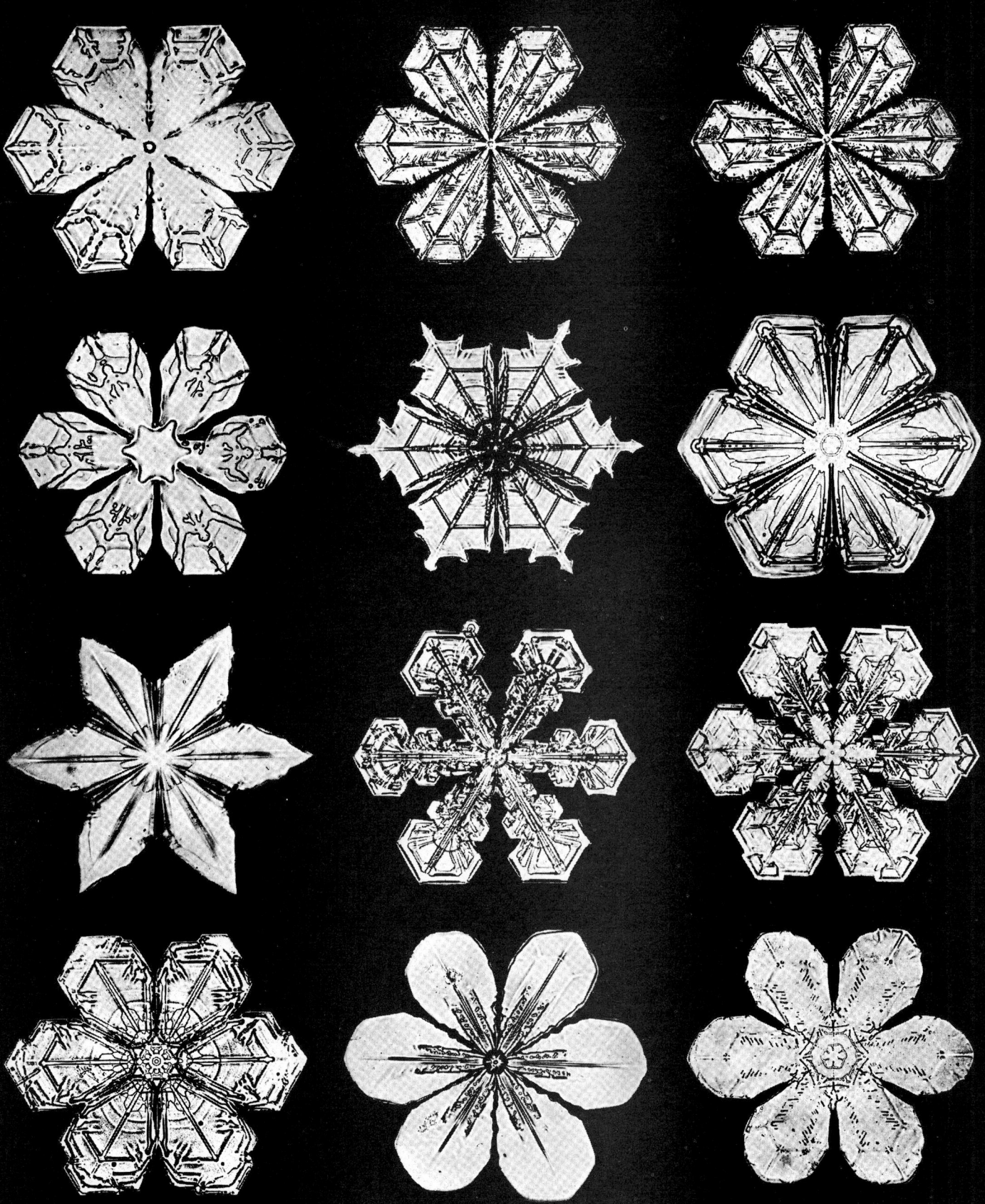

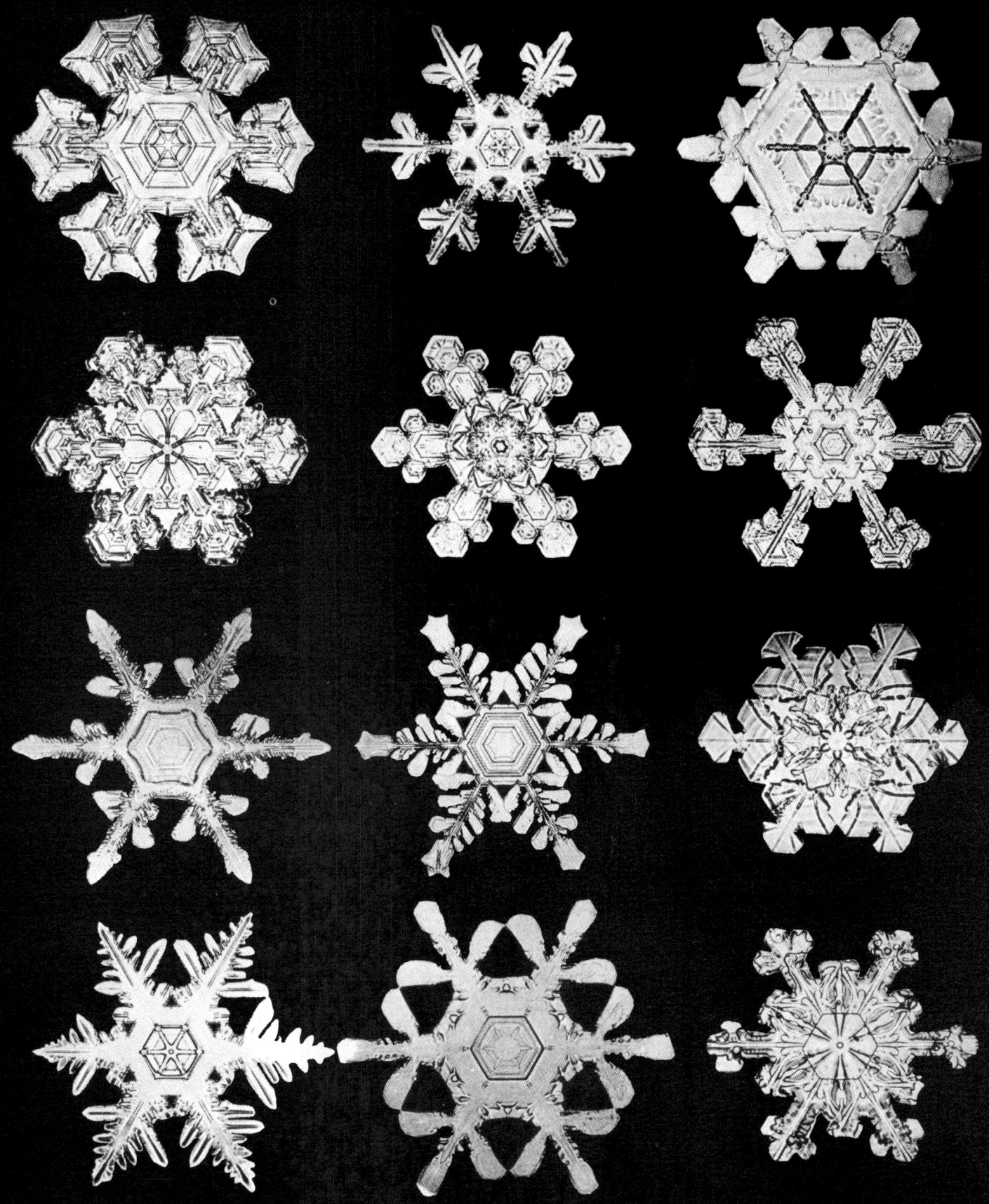

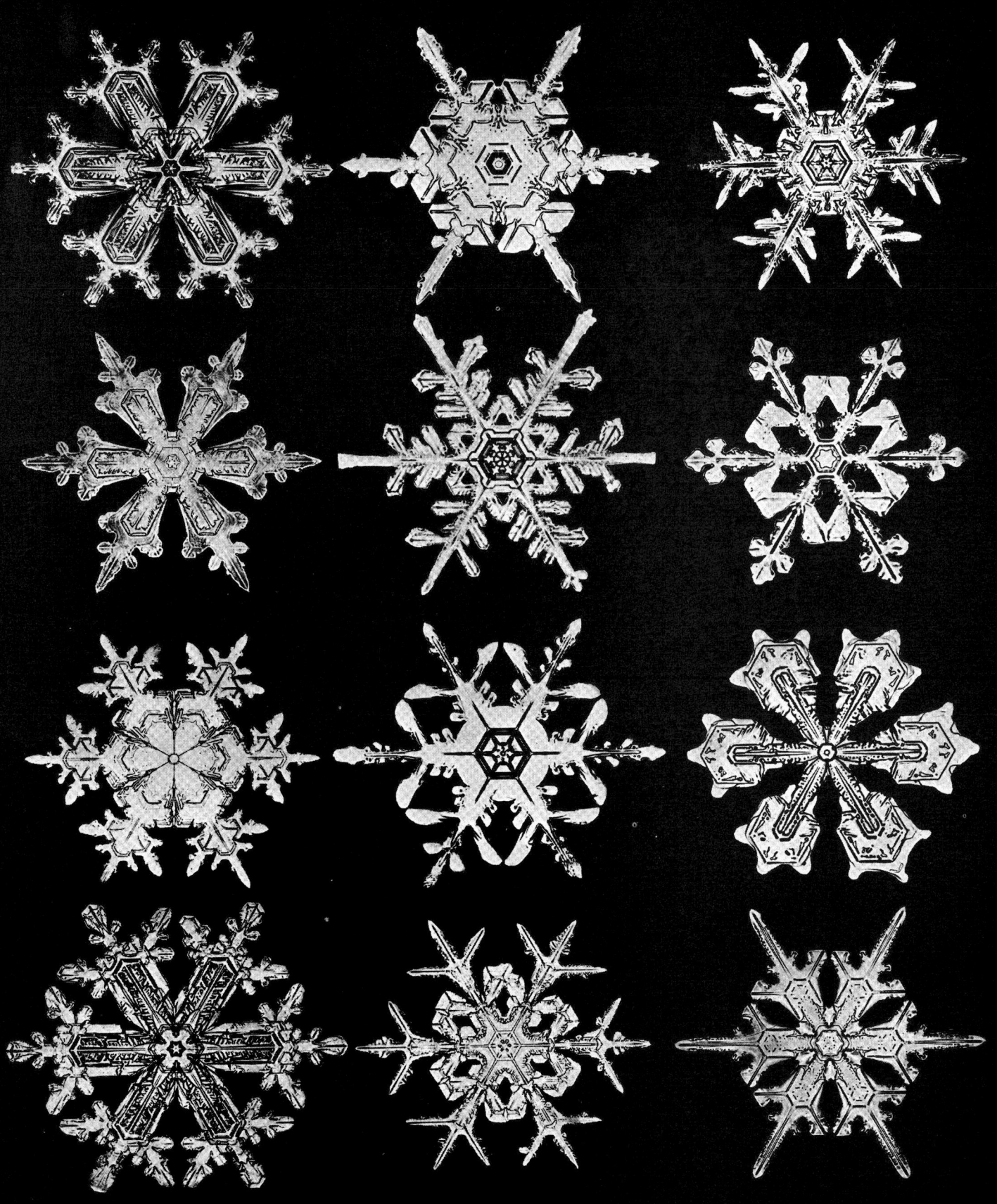

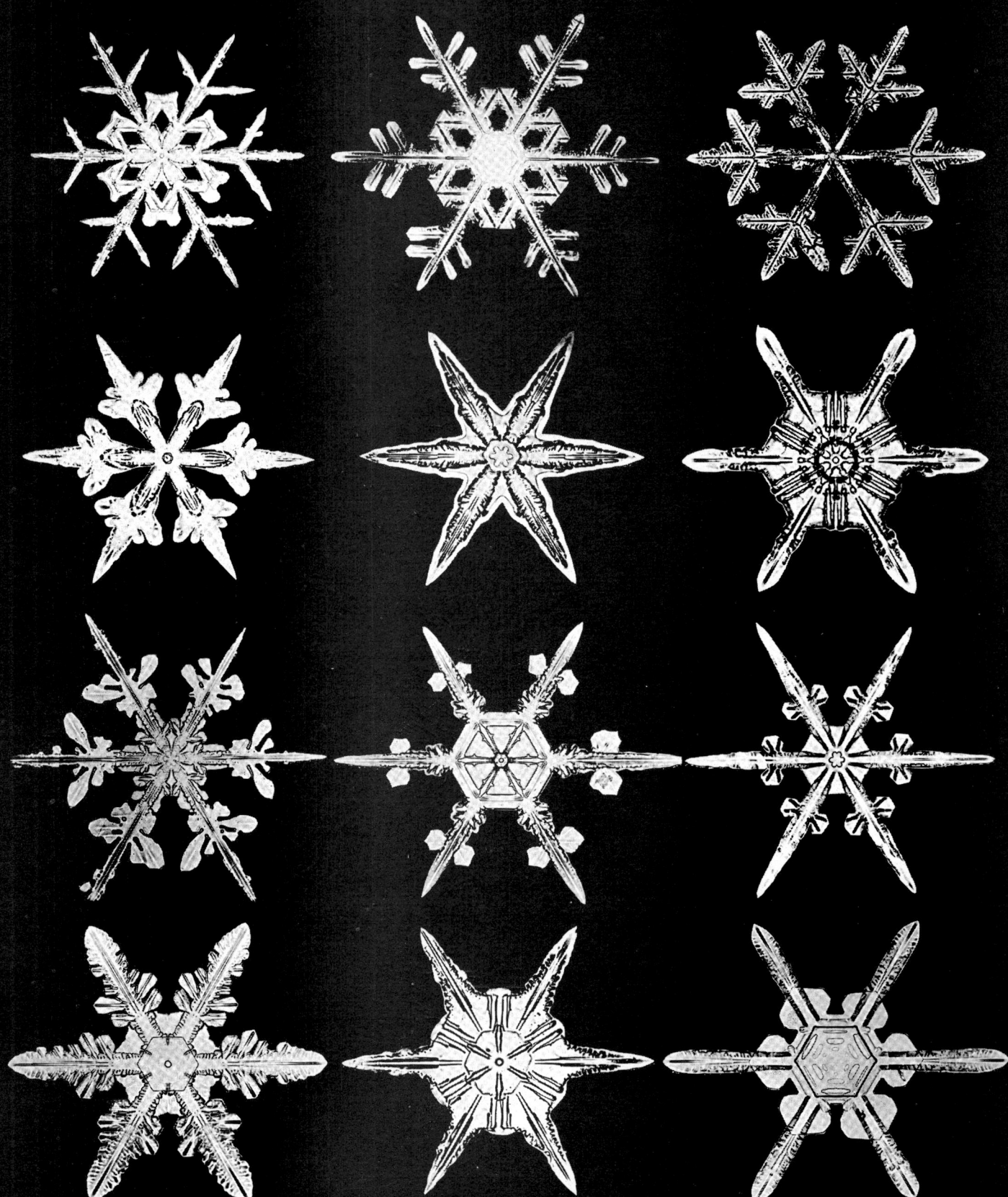

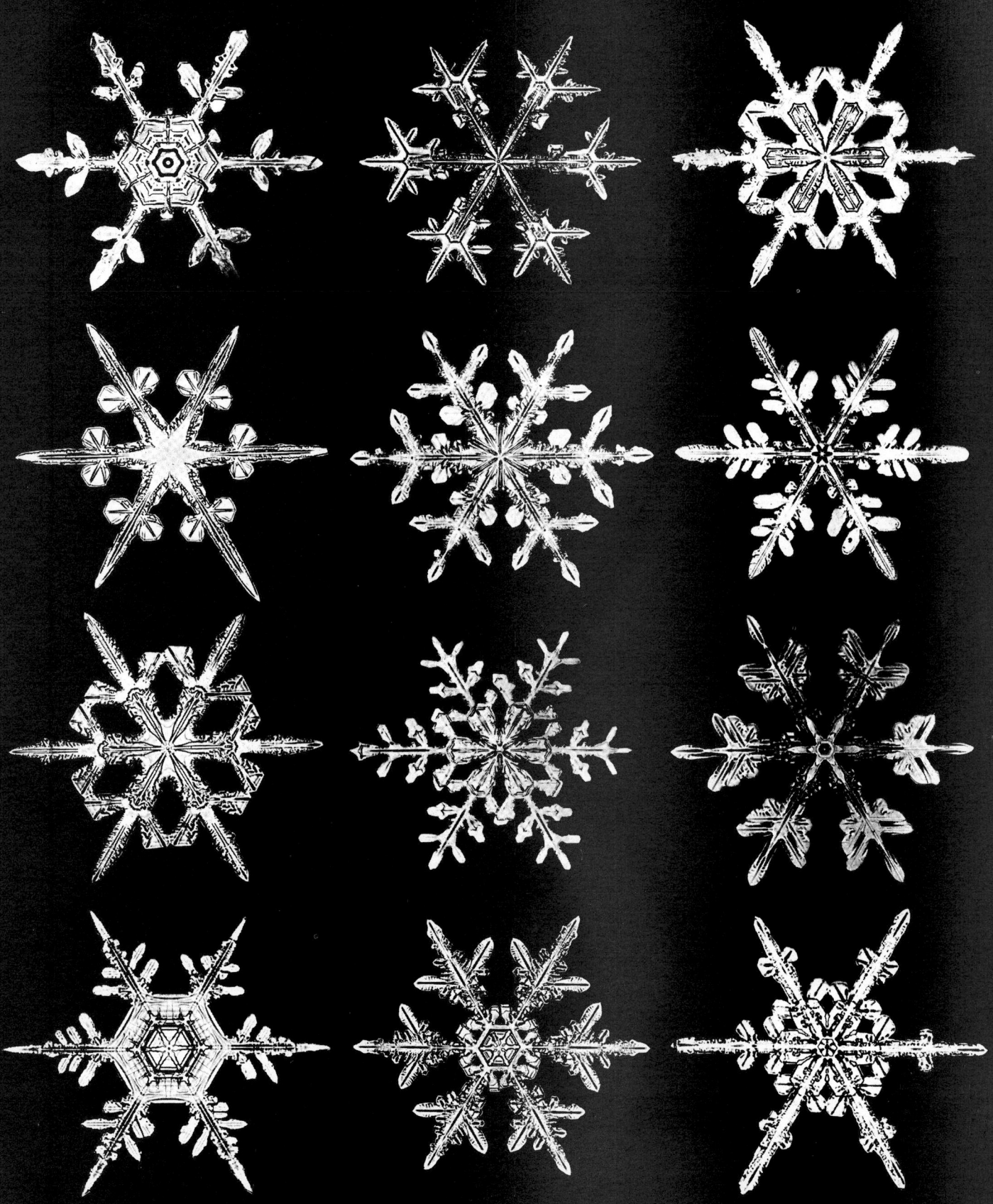

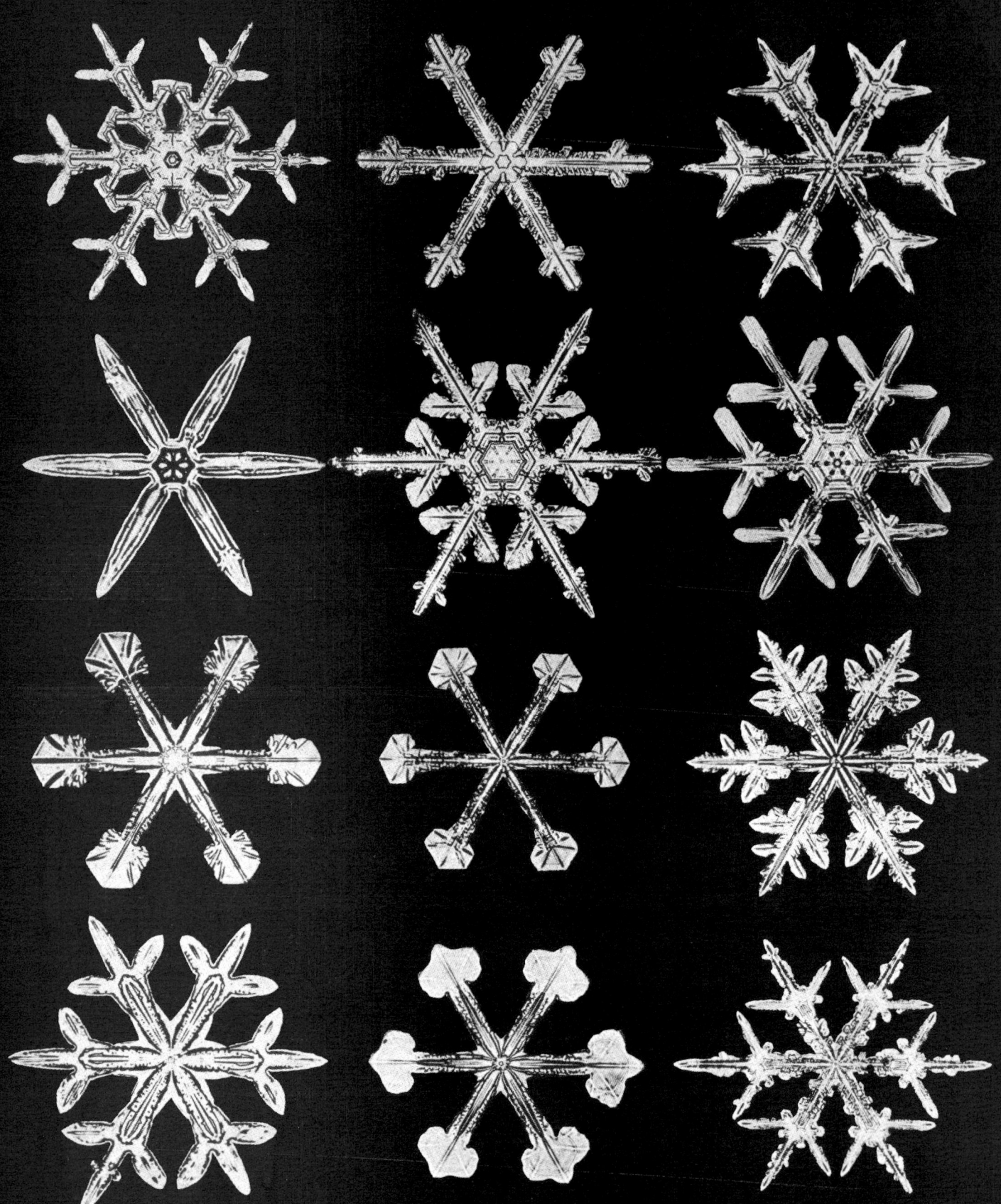

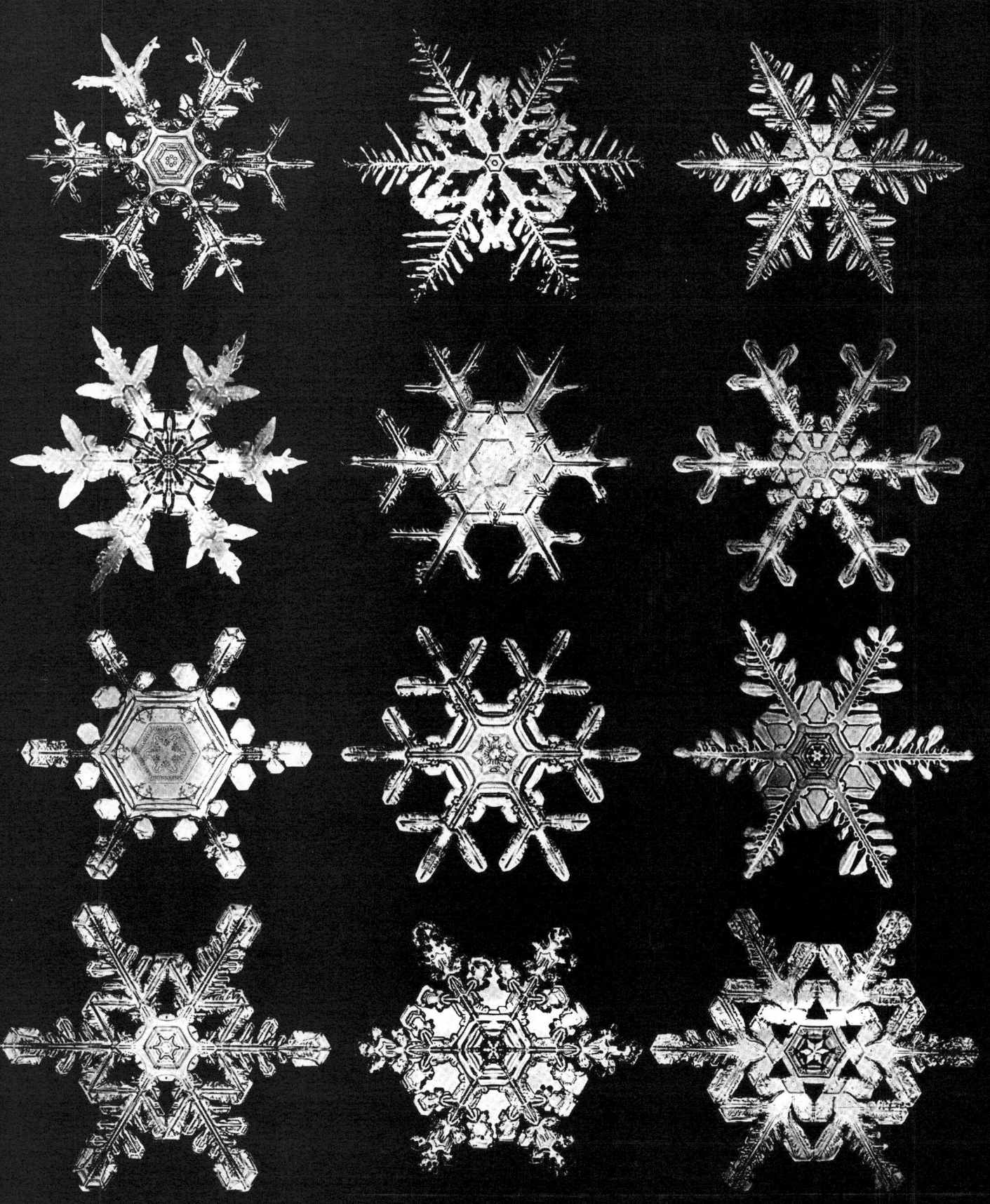

41

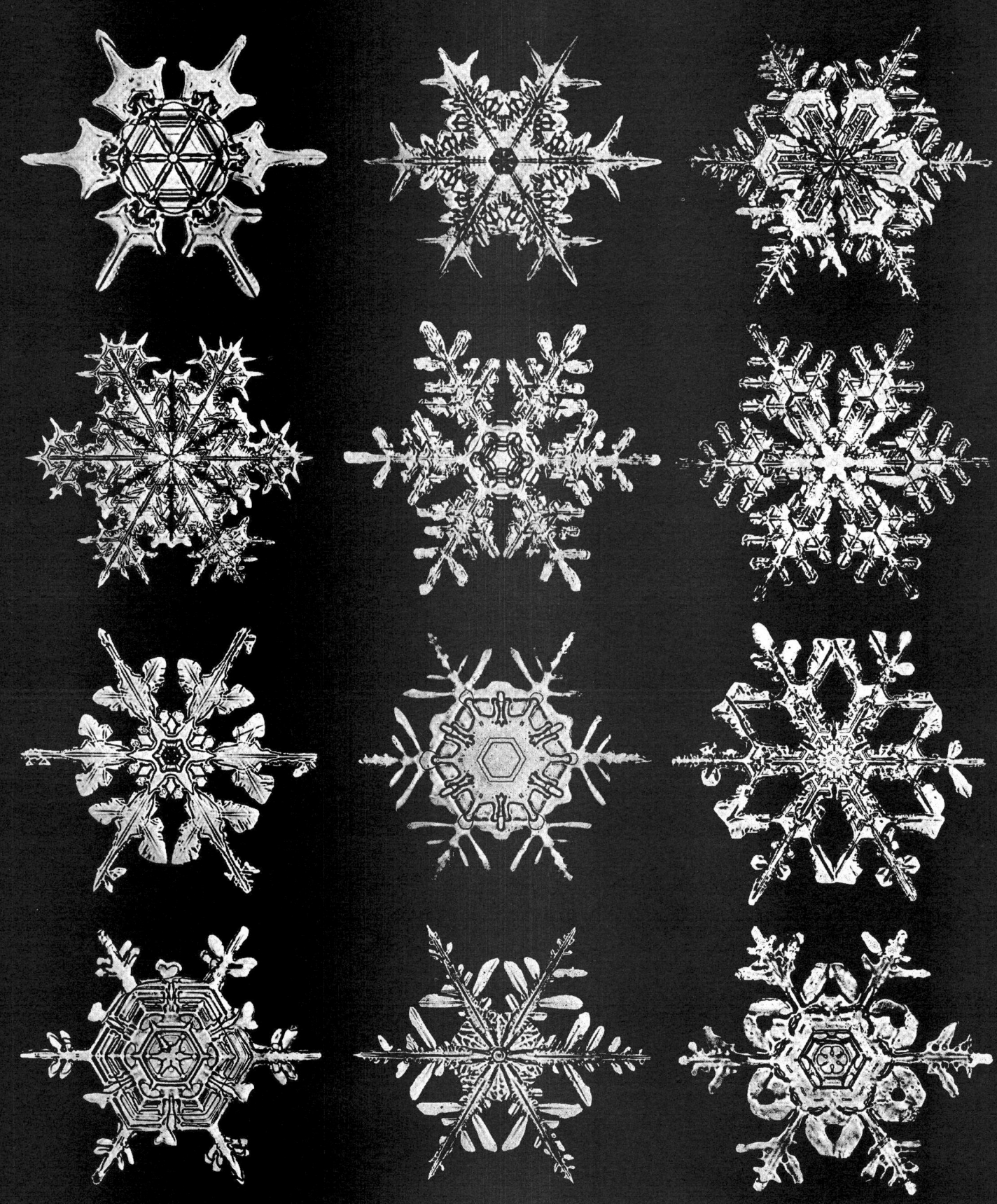

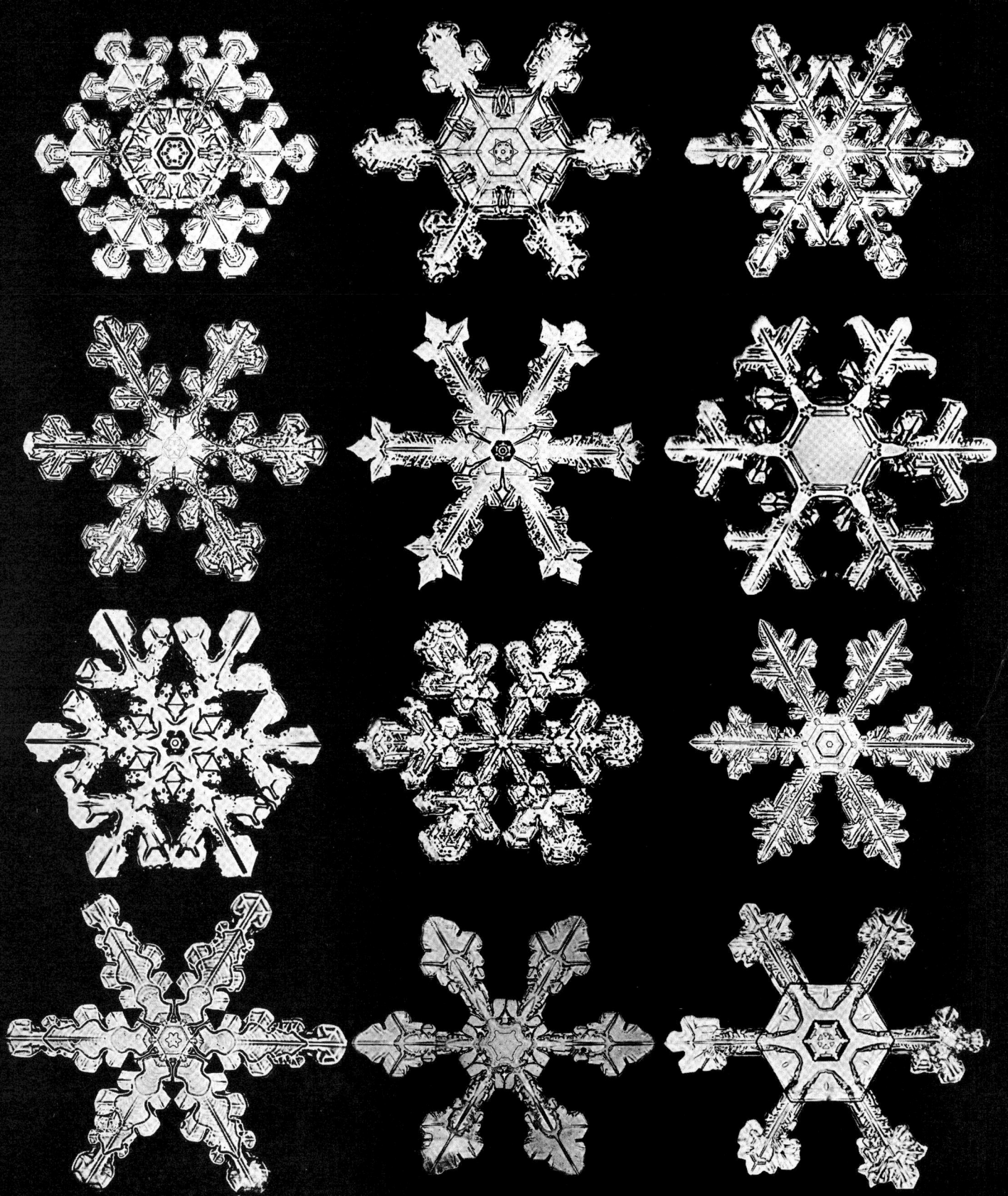

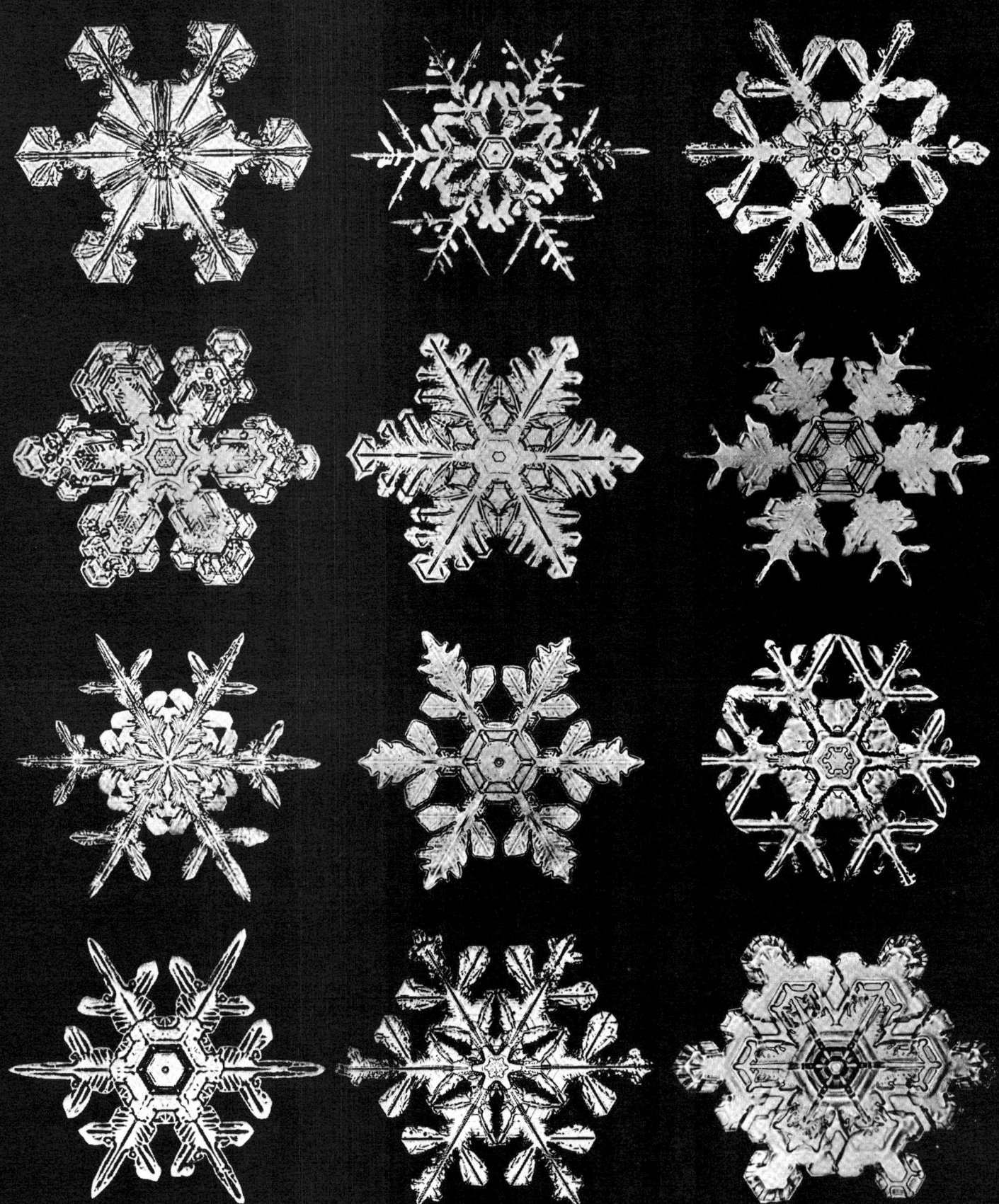

44

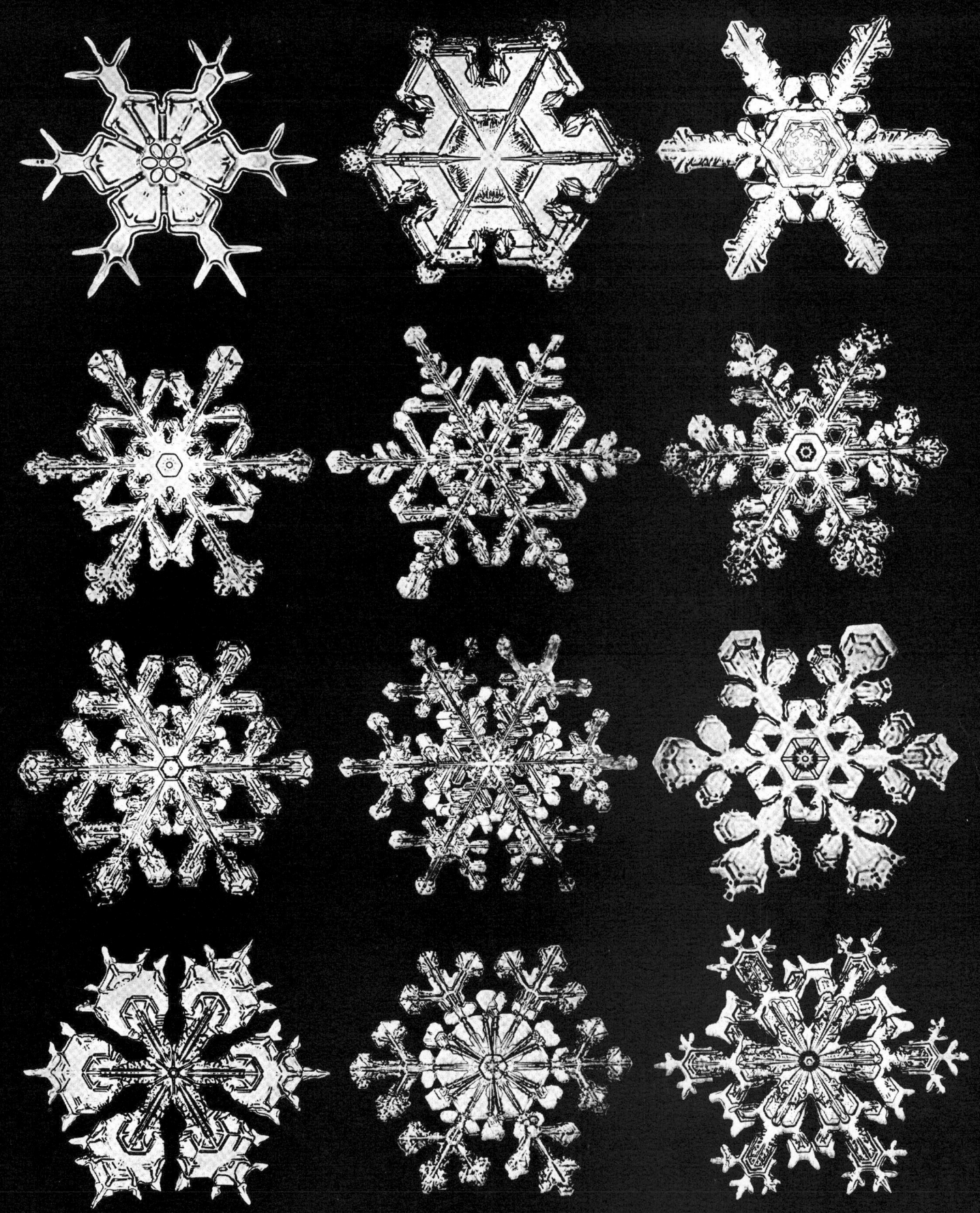

45

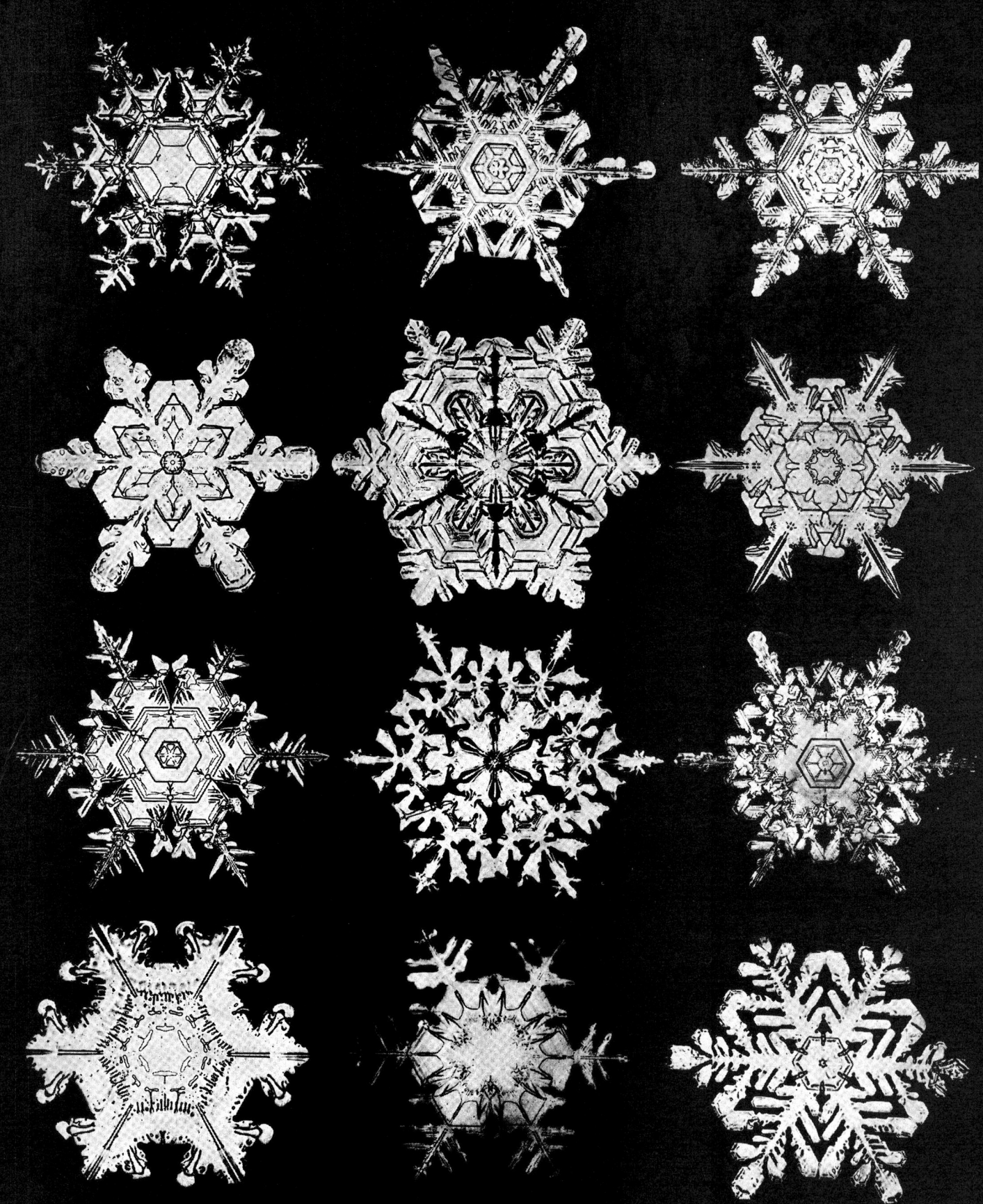

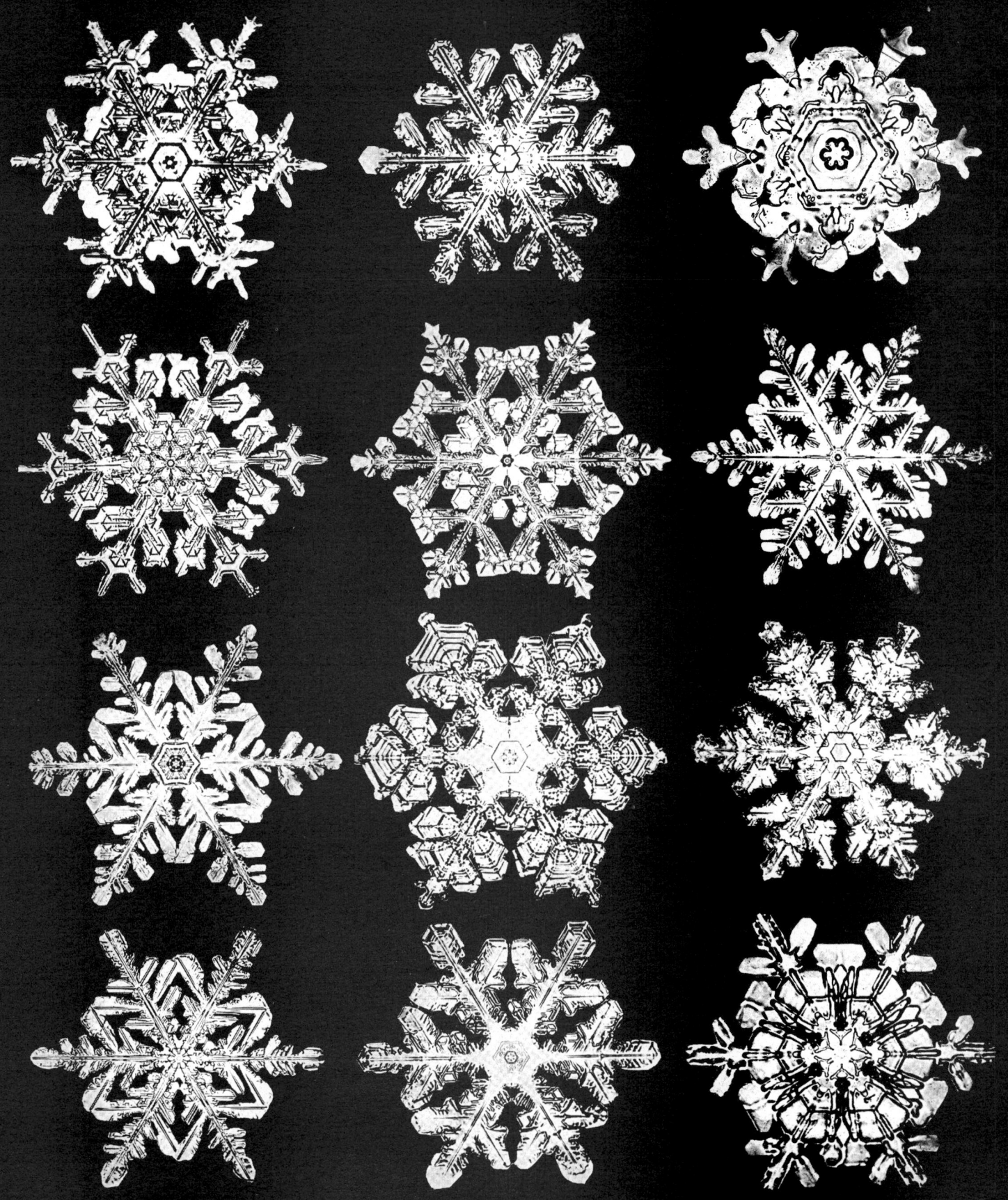

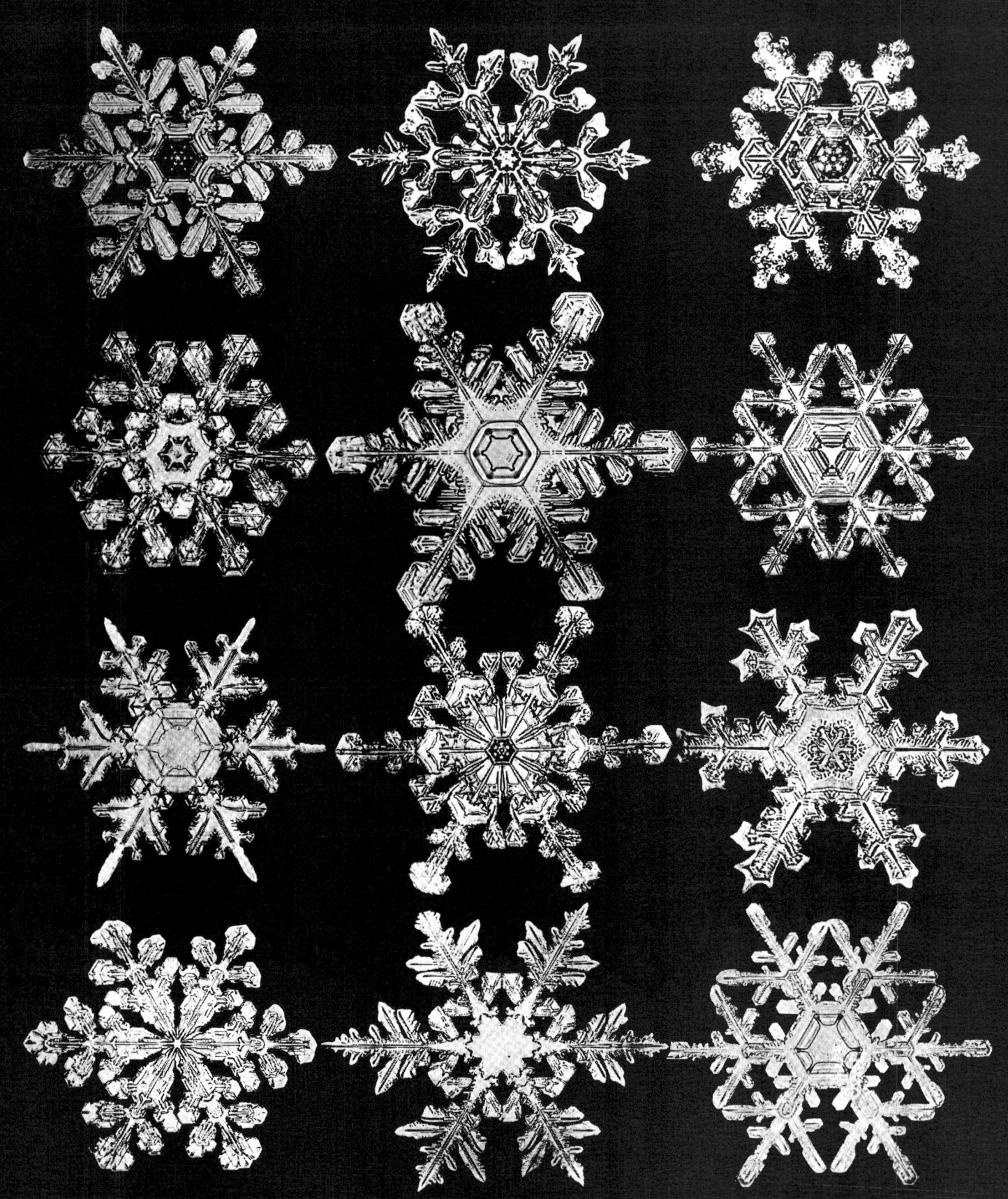

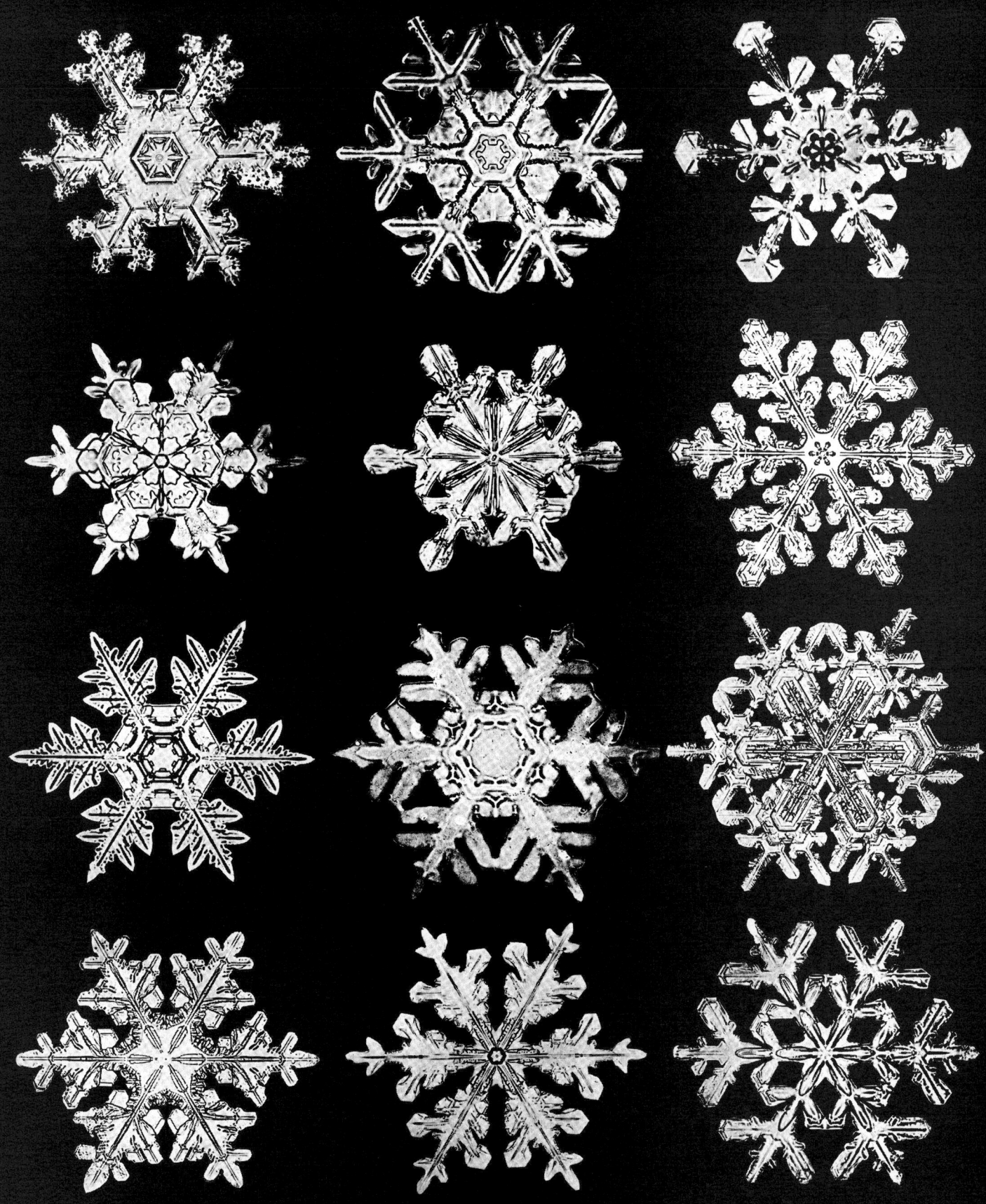

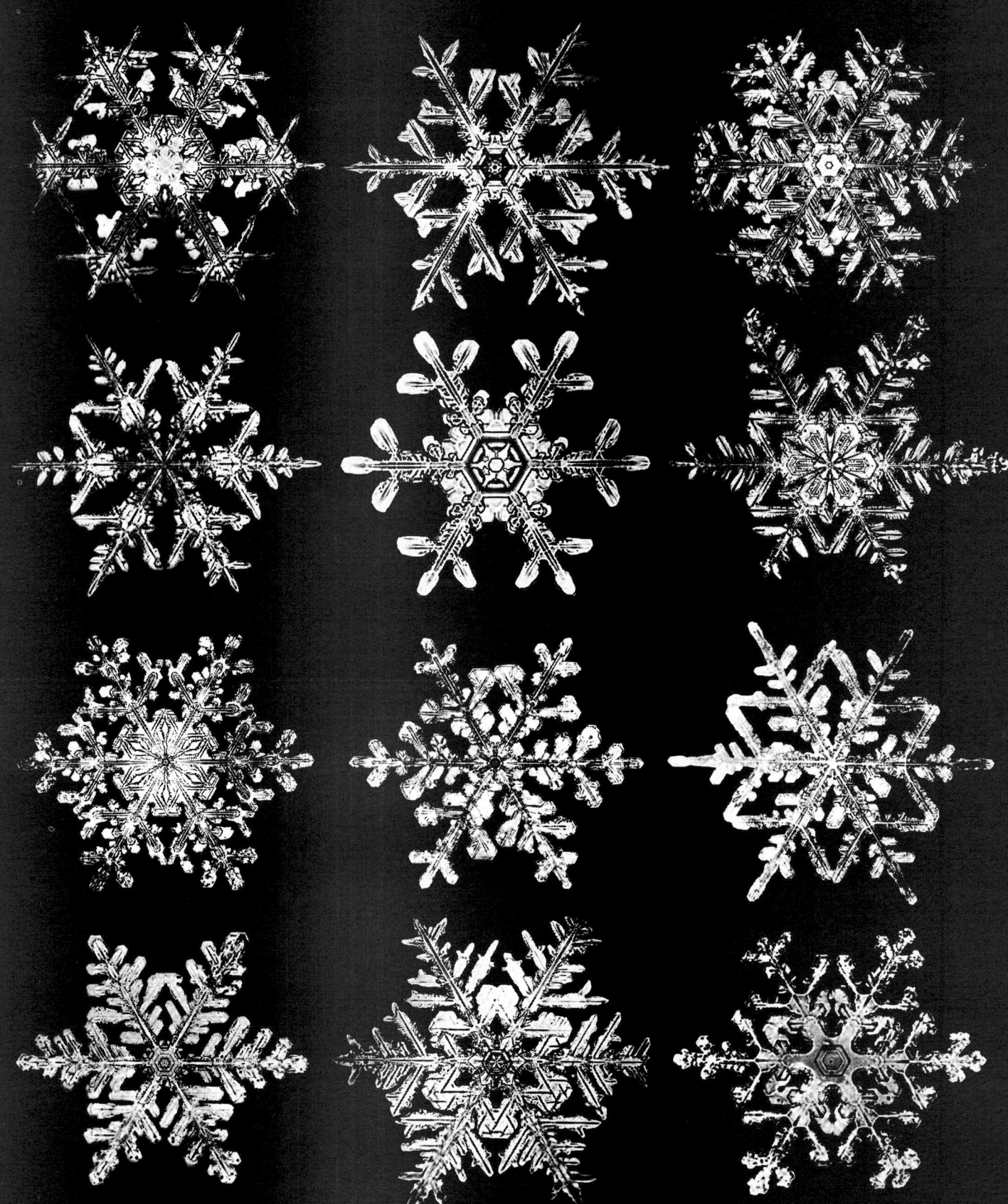

51

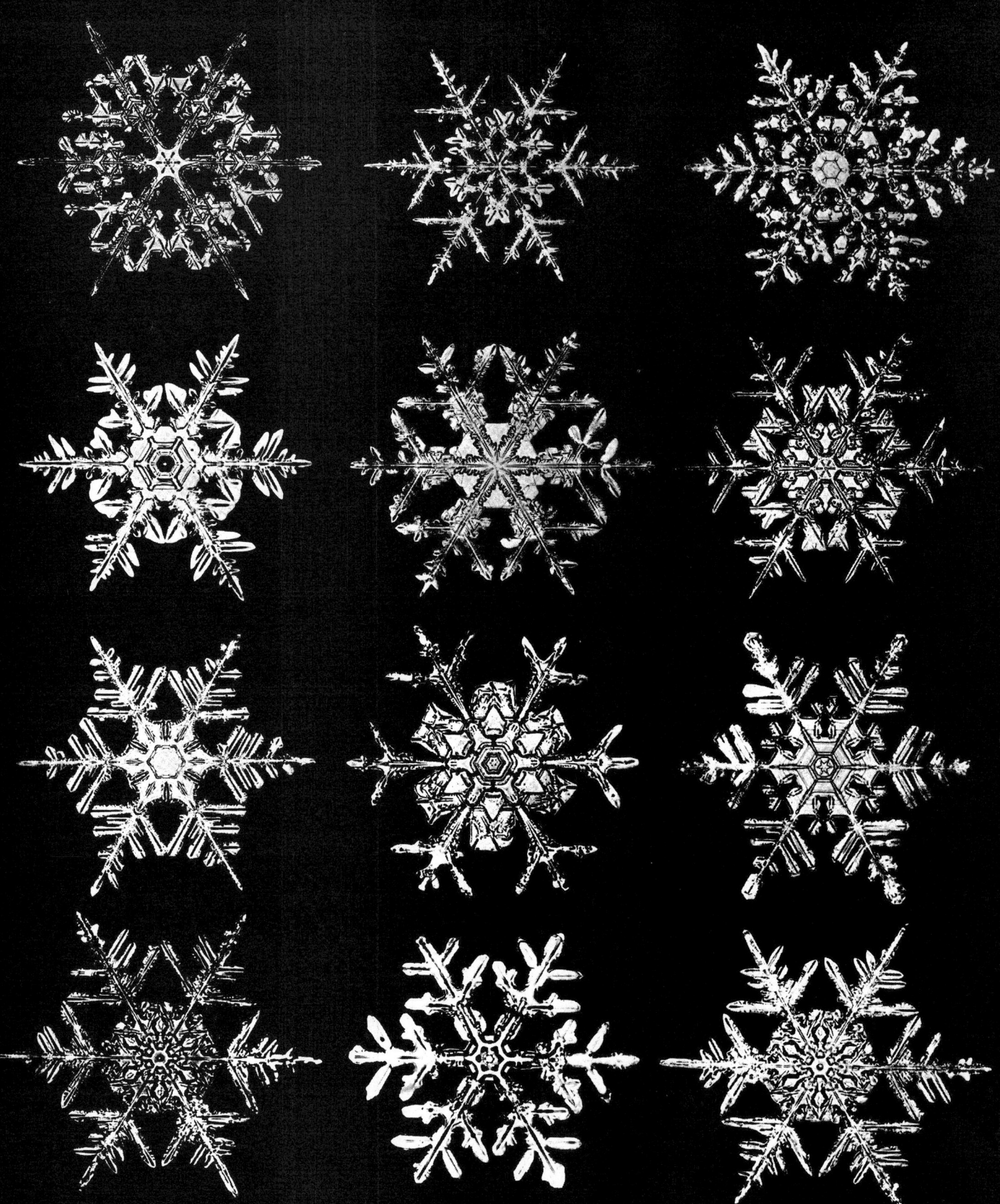

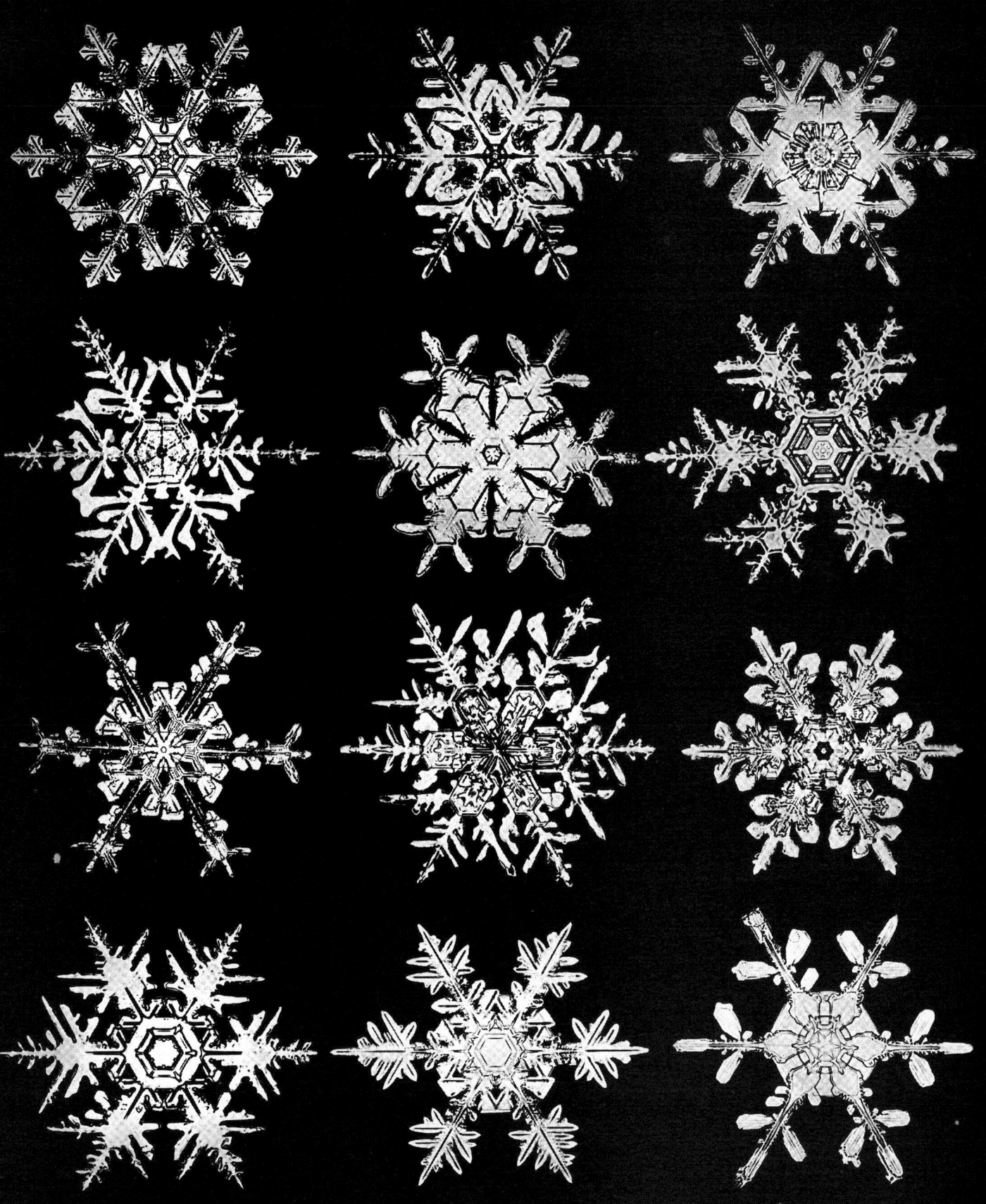

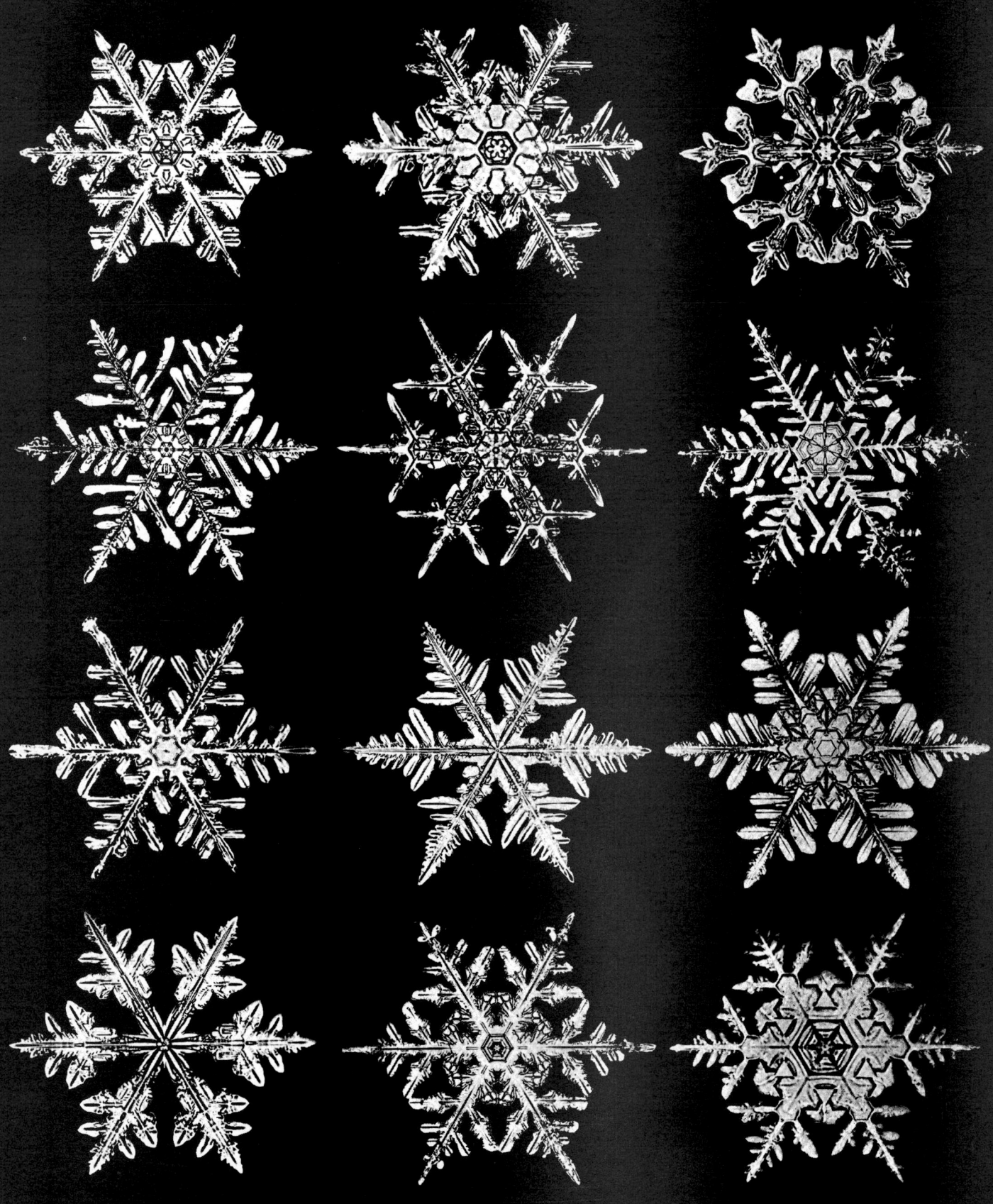

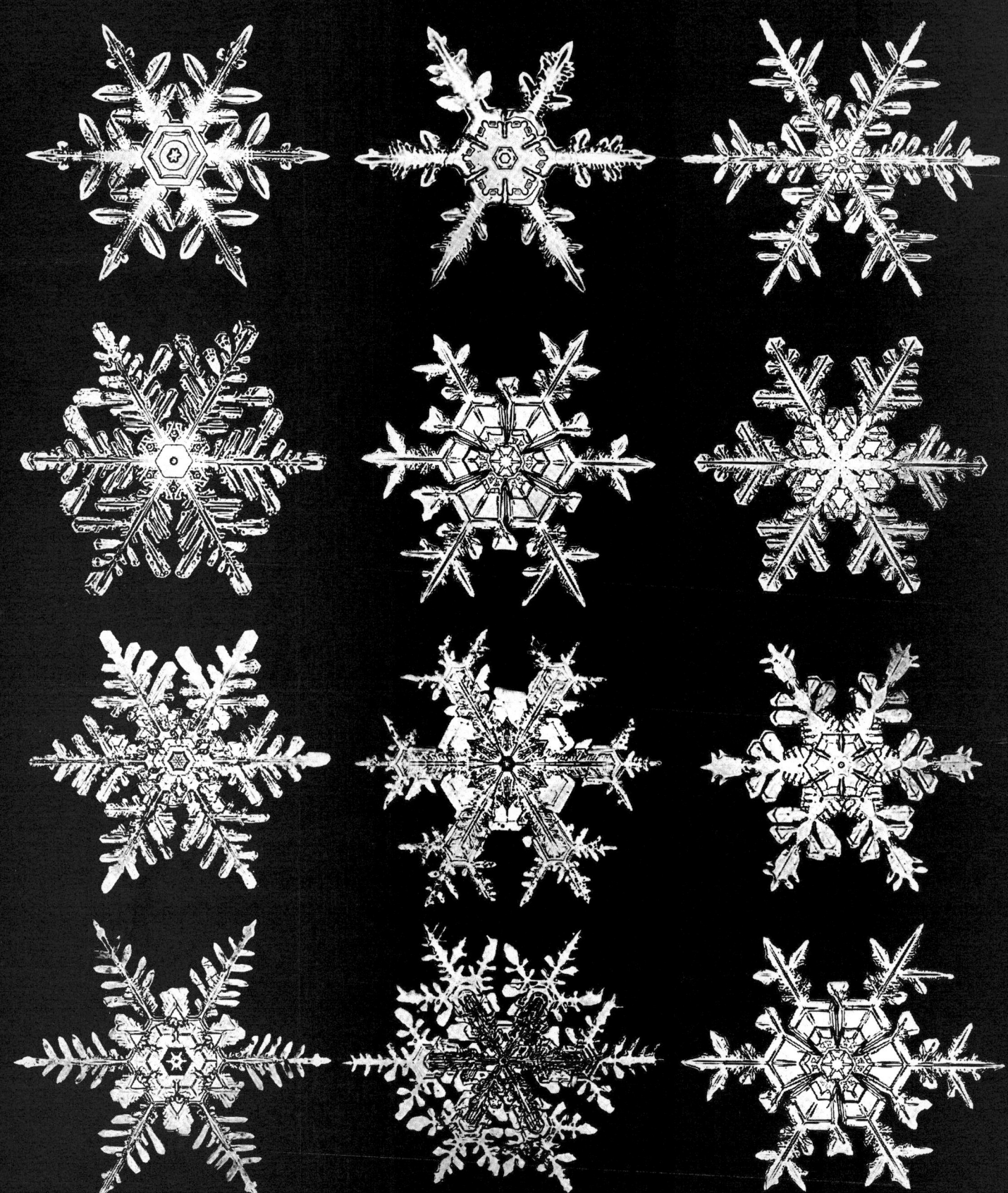

56

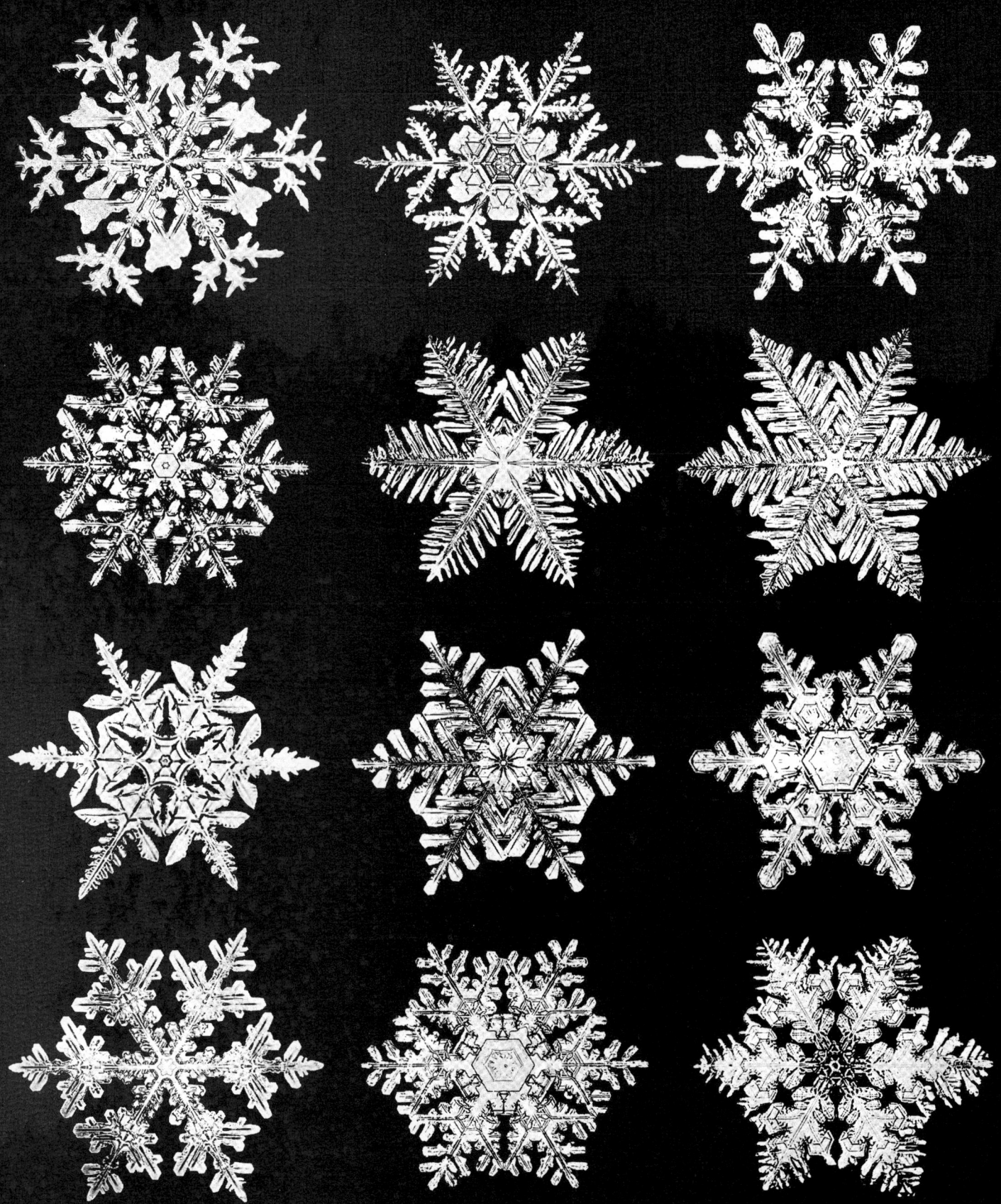

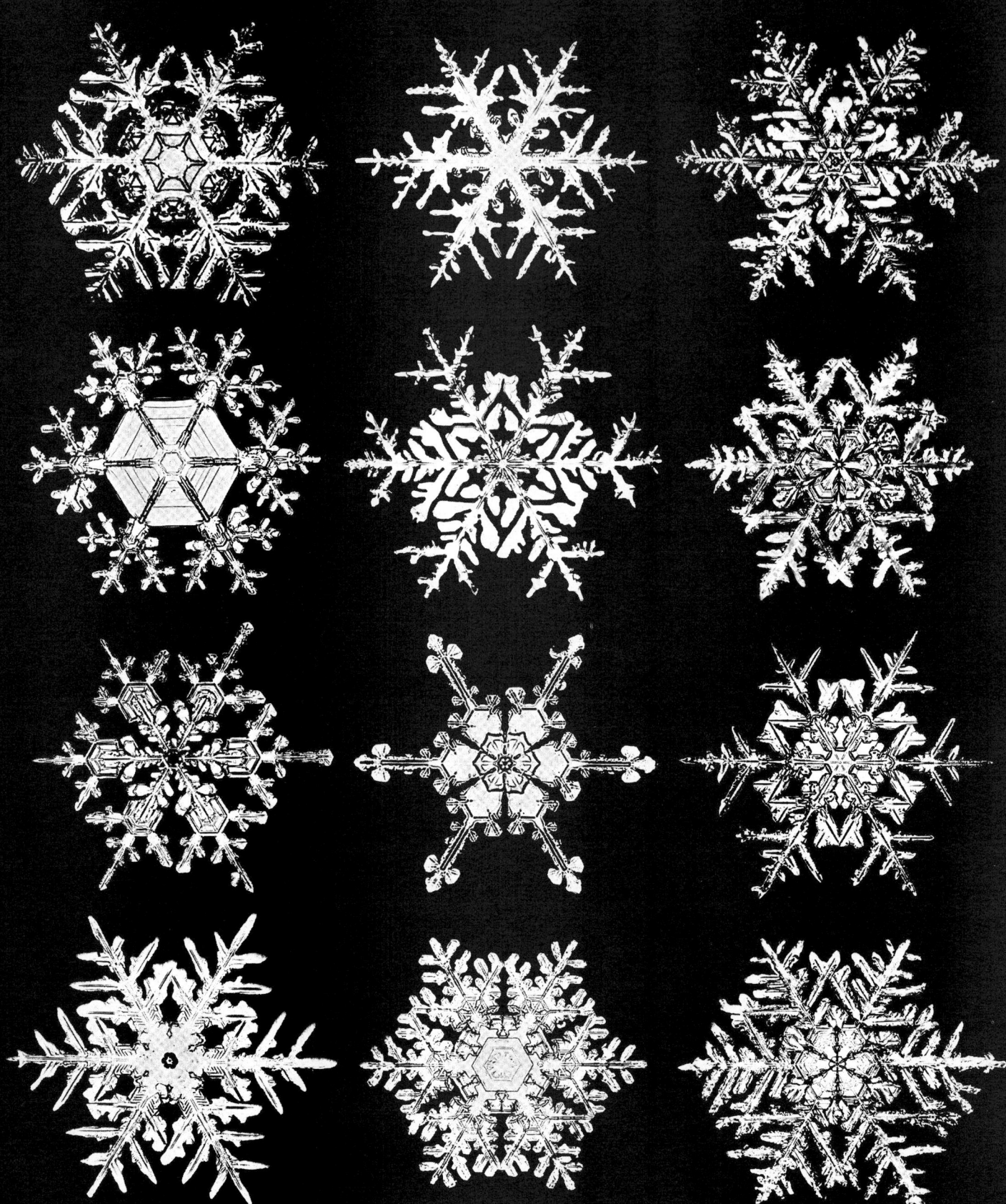

58

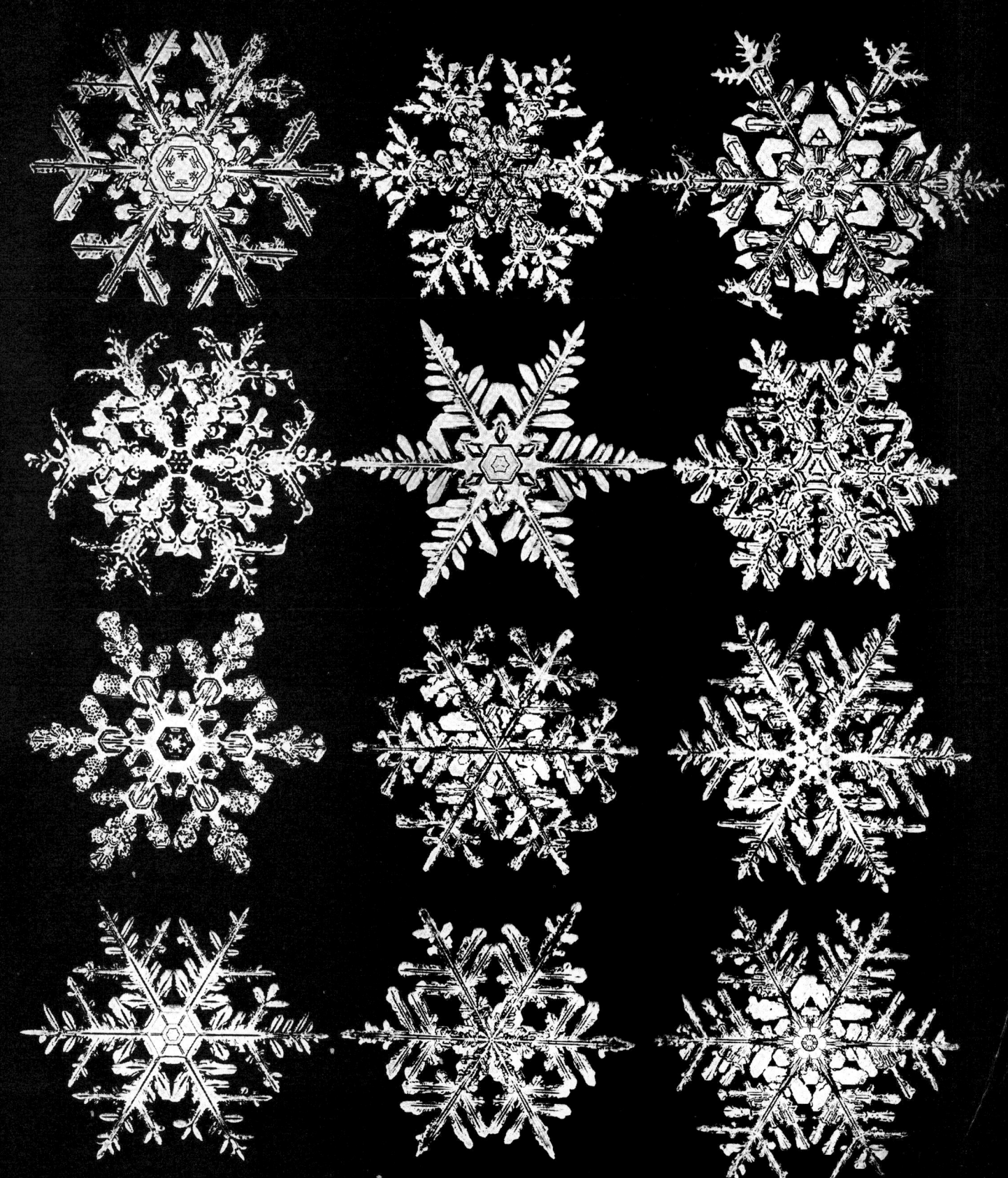

59

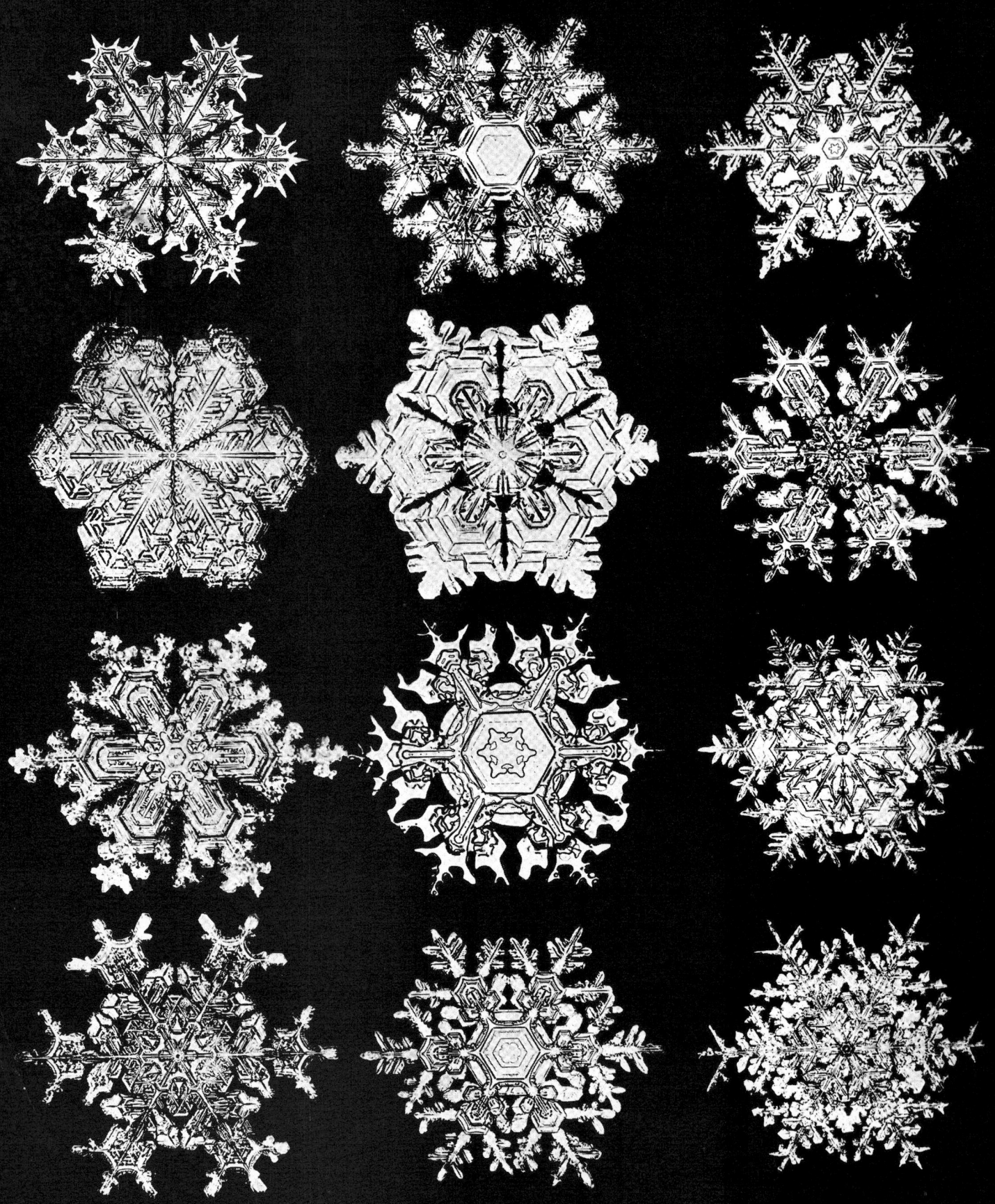

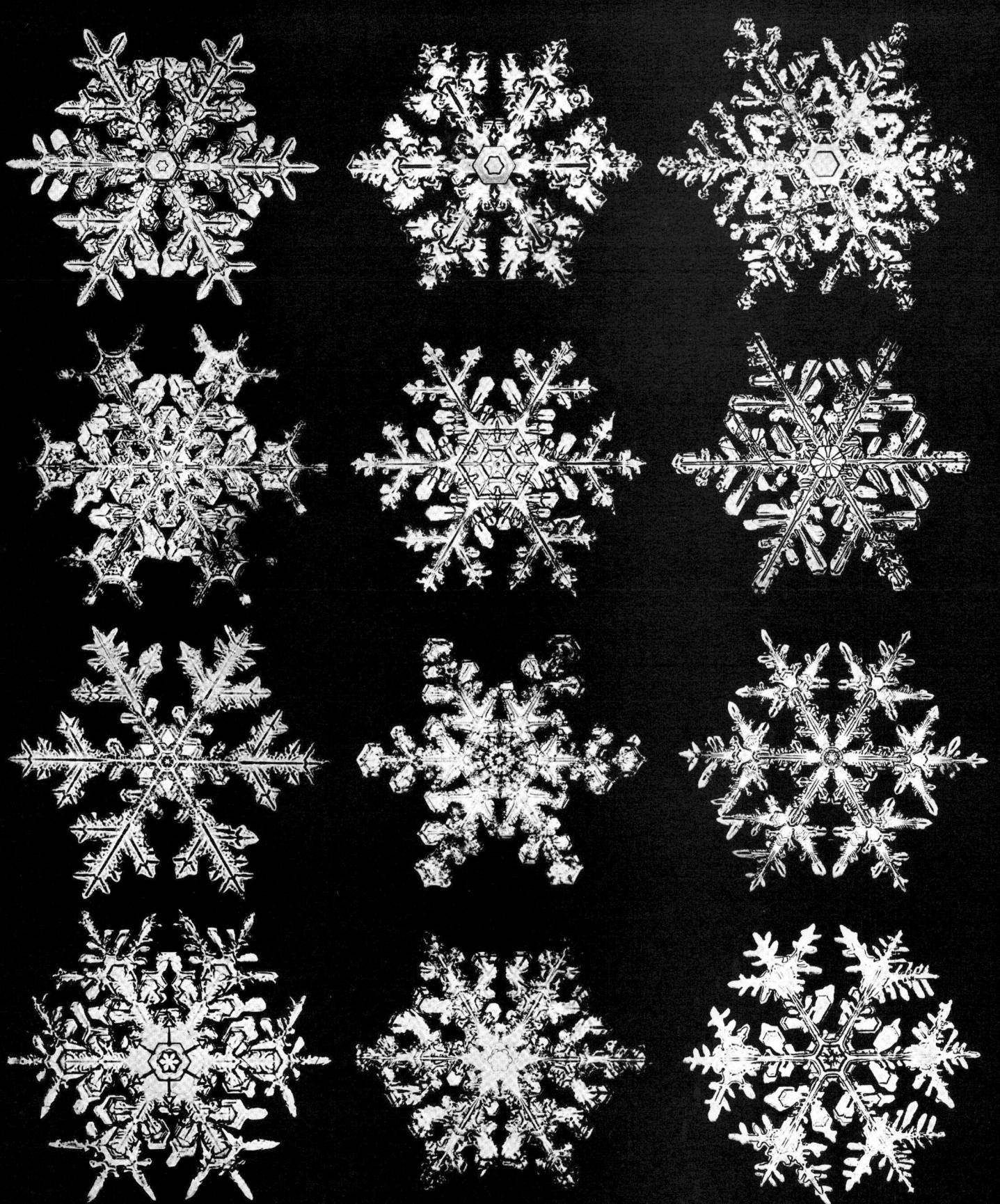

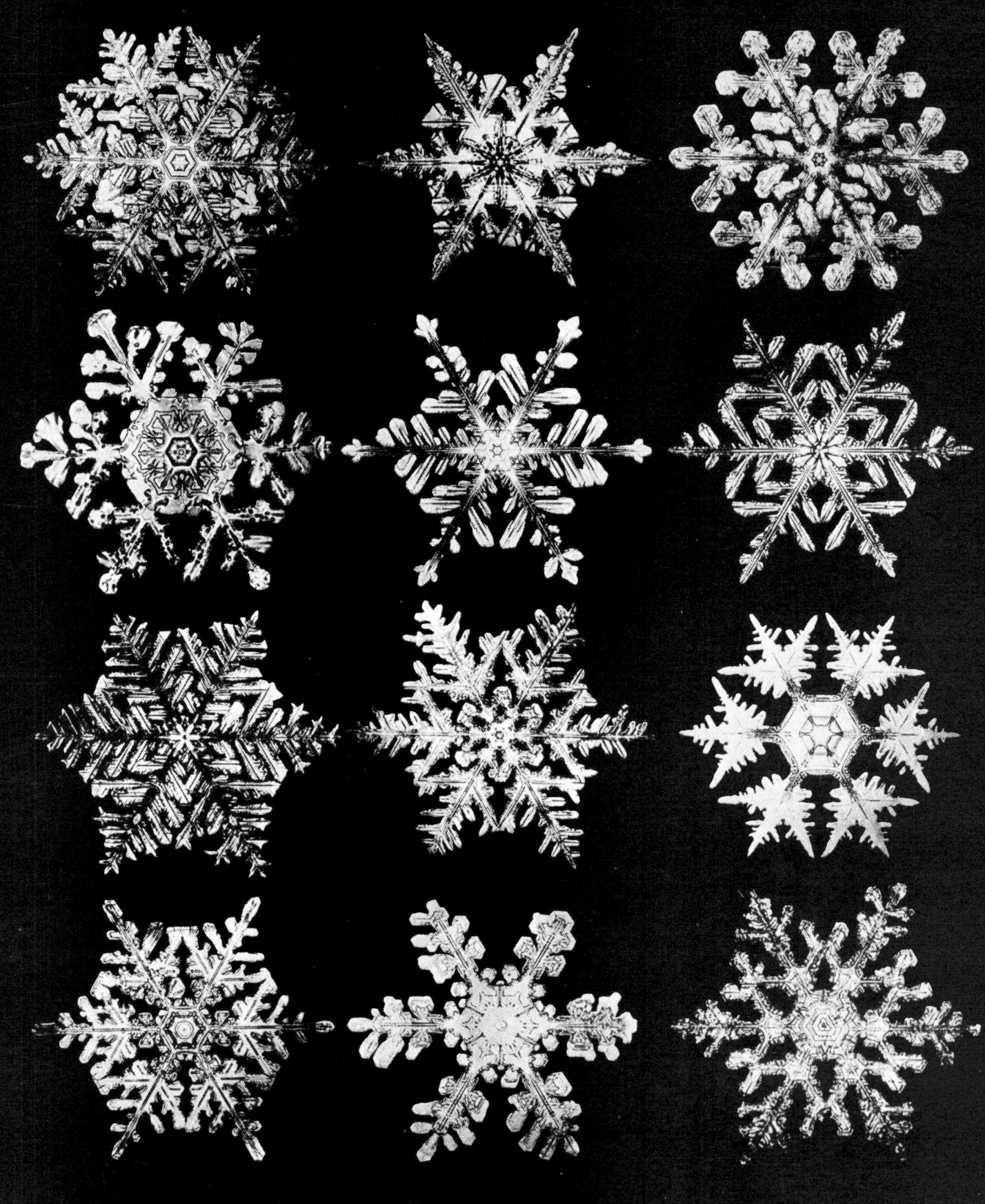

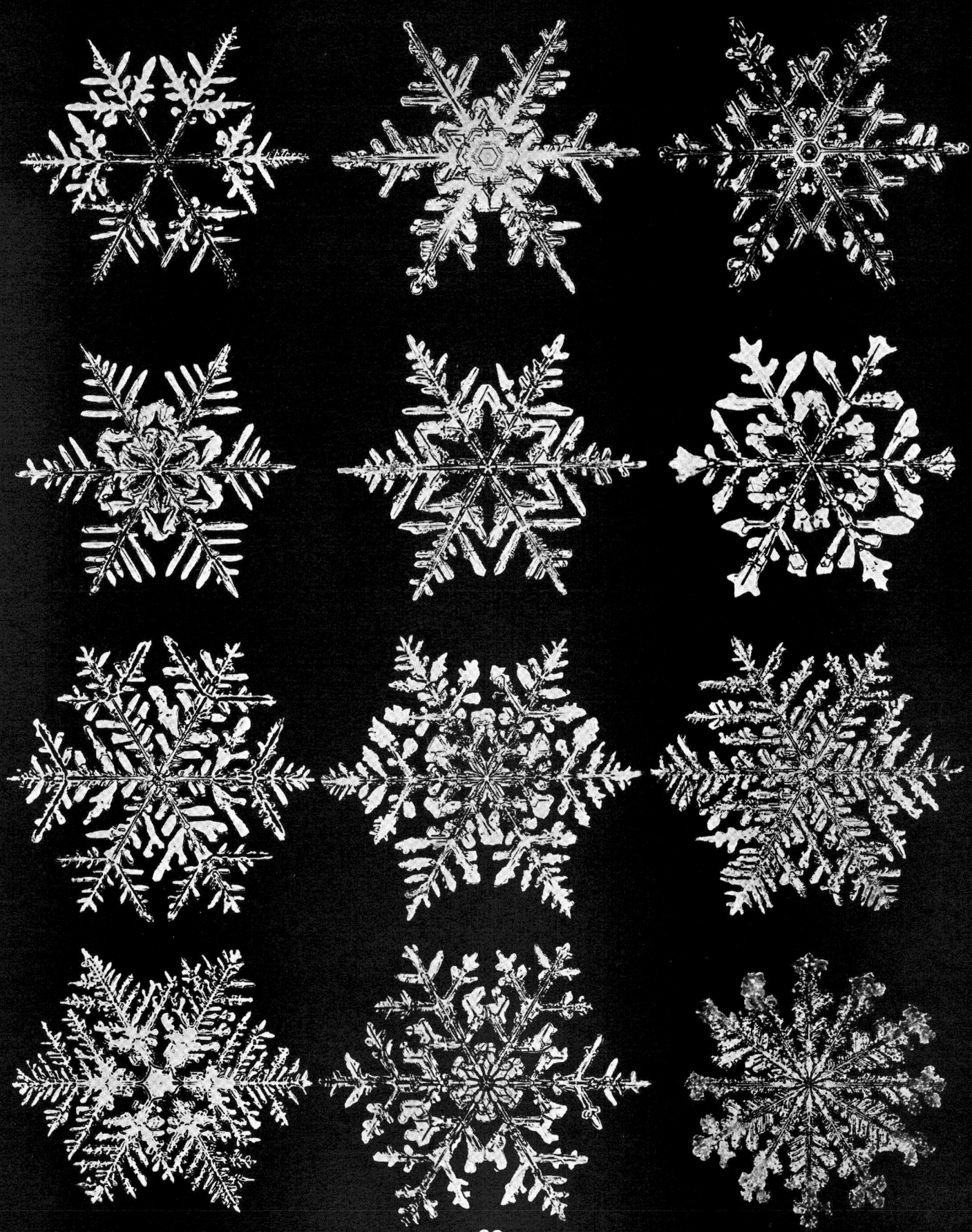

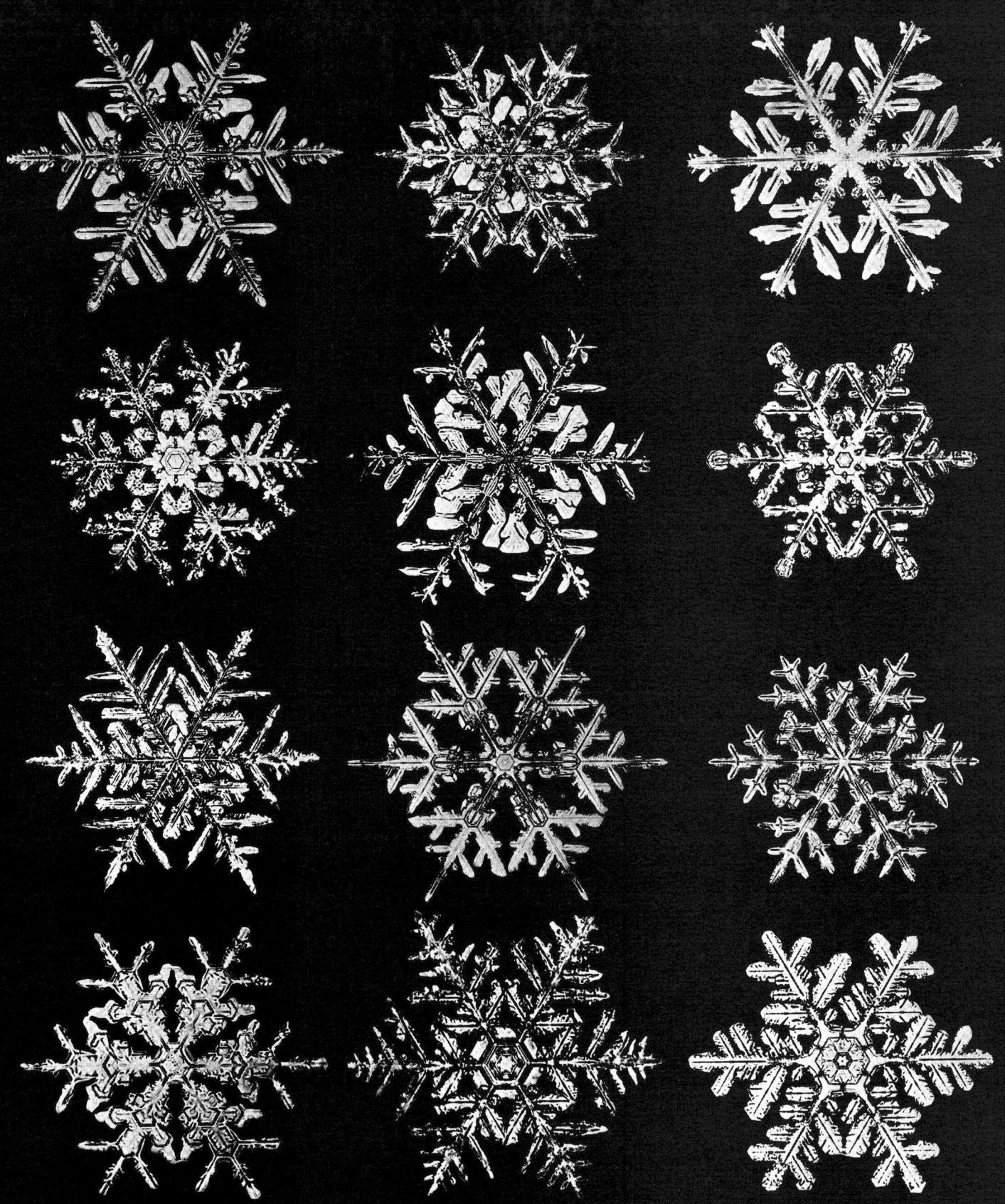

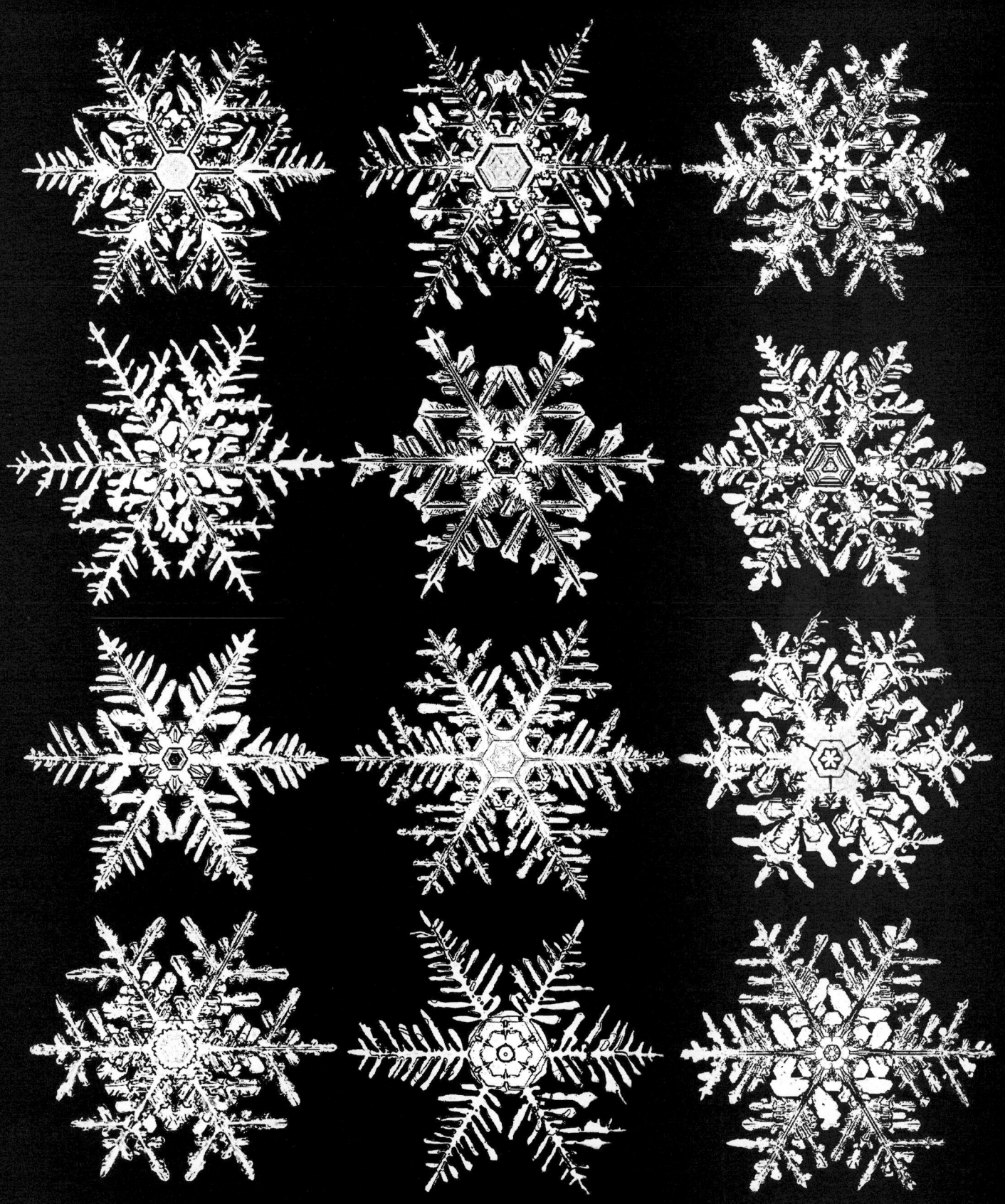

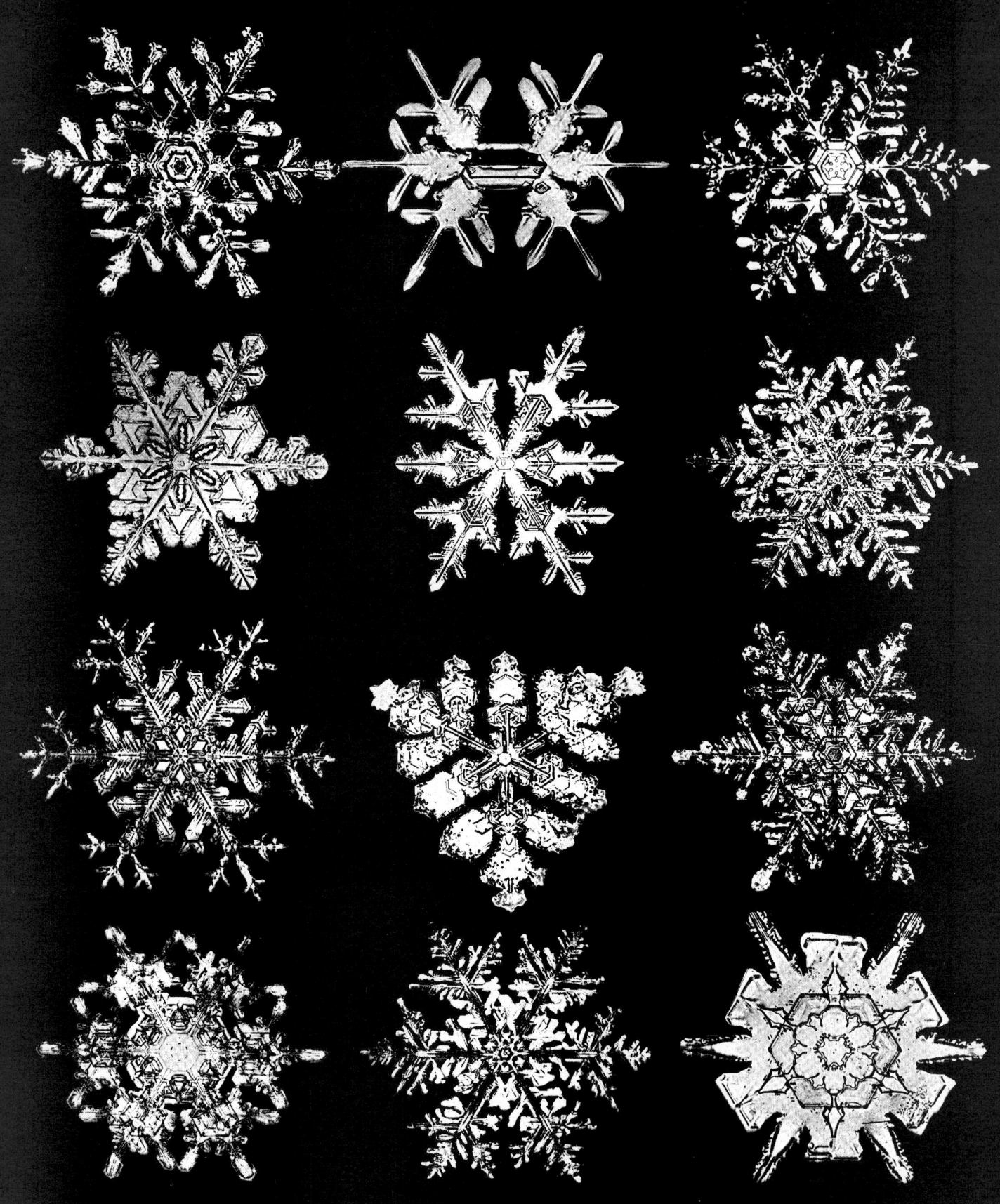

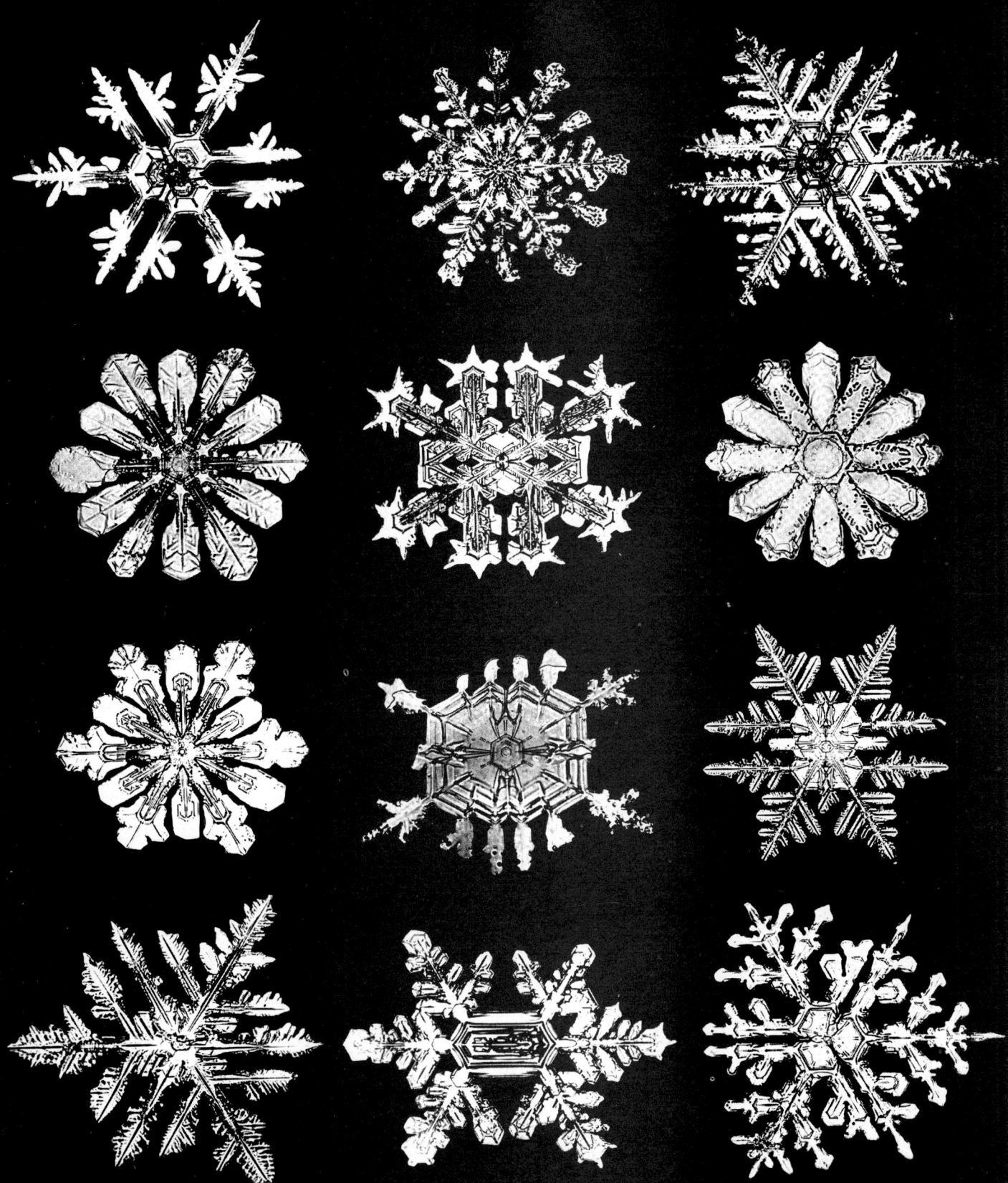

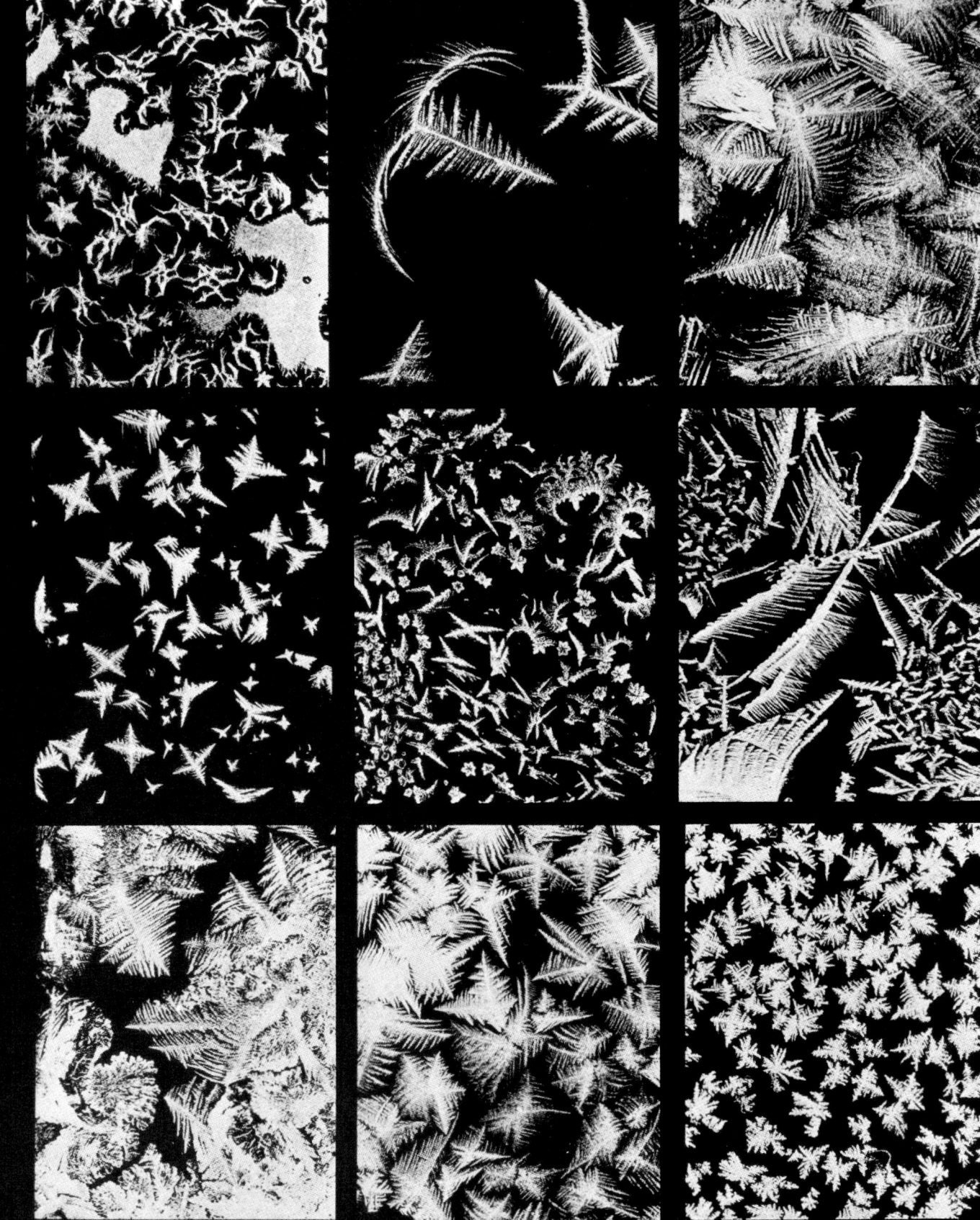

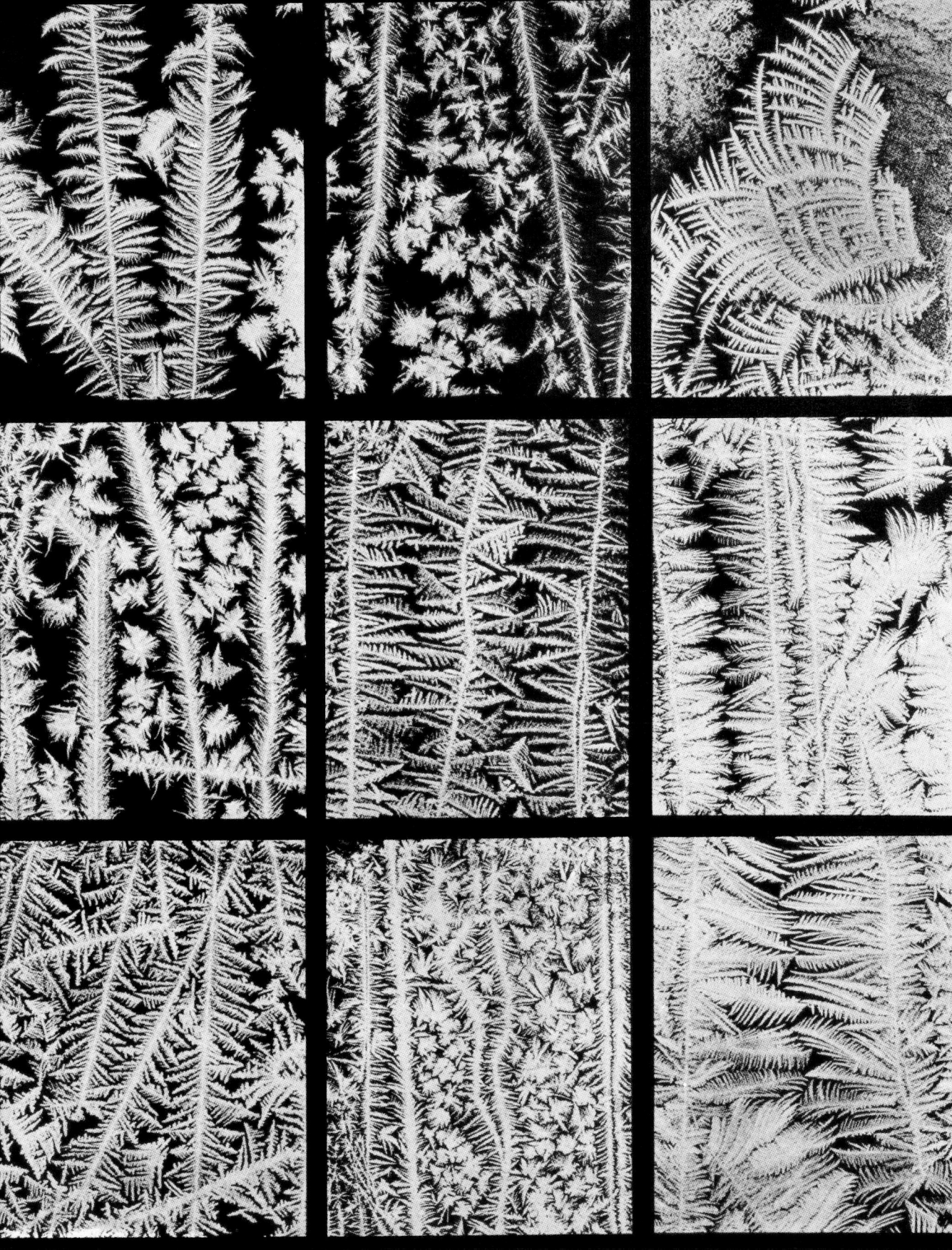

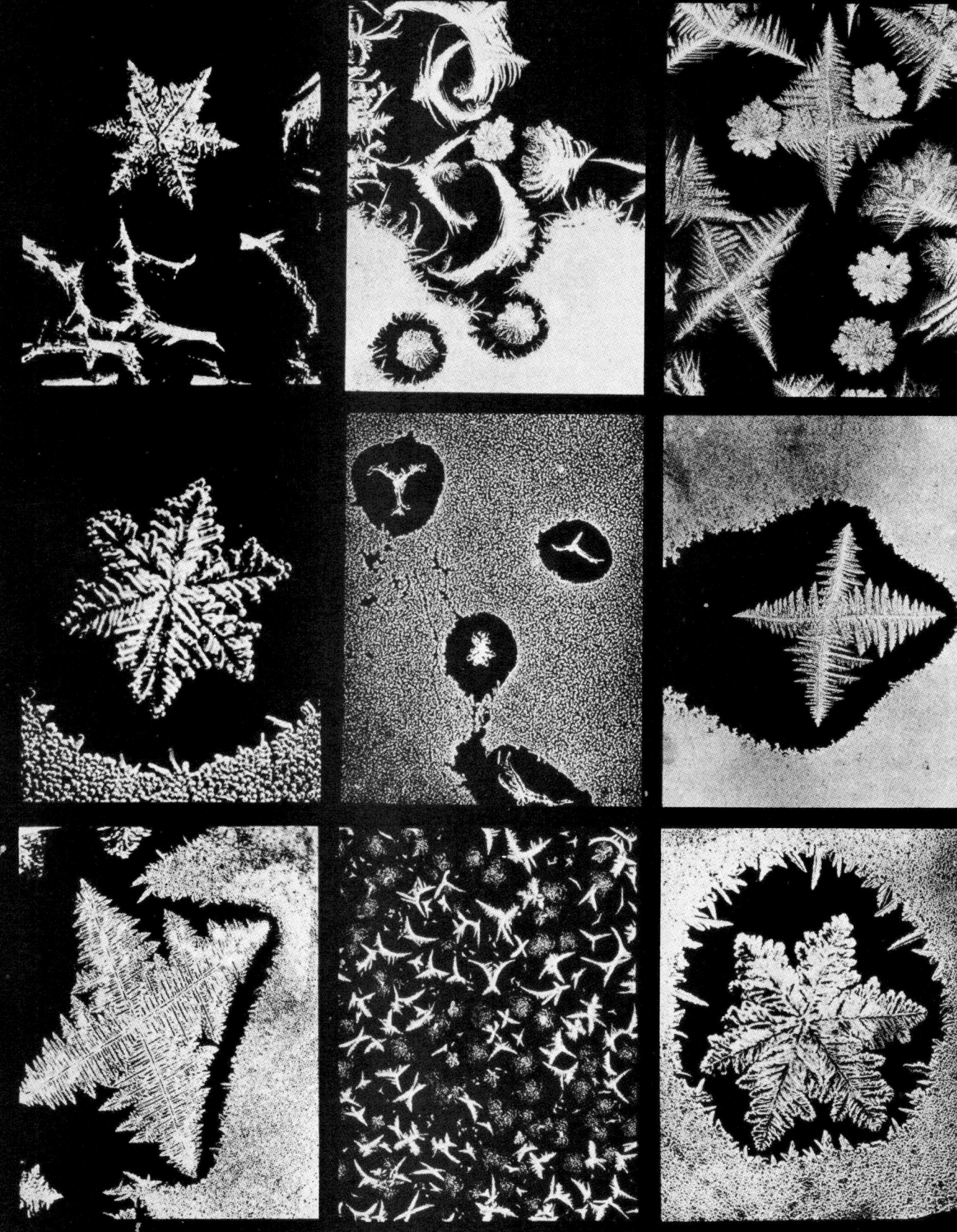

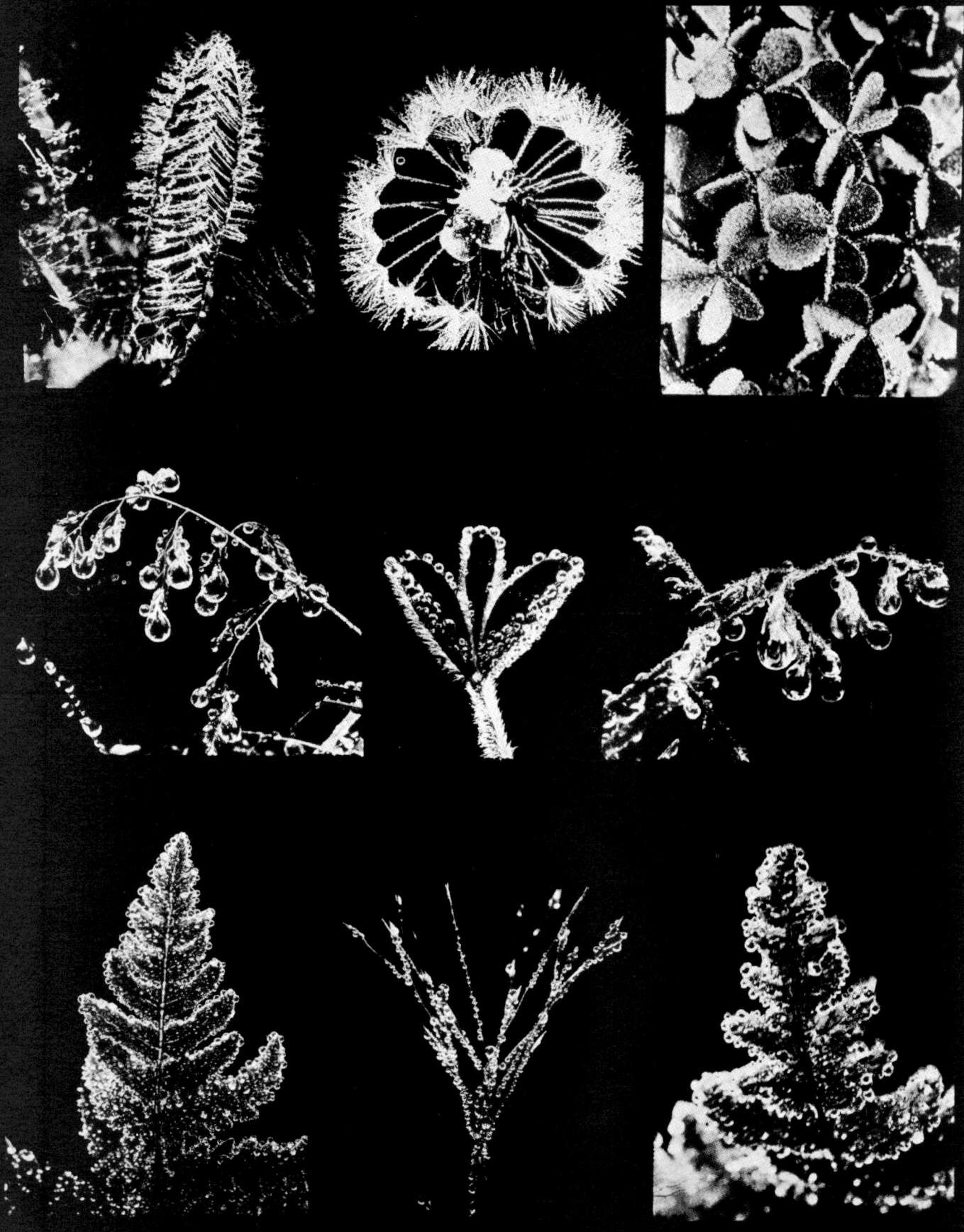

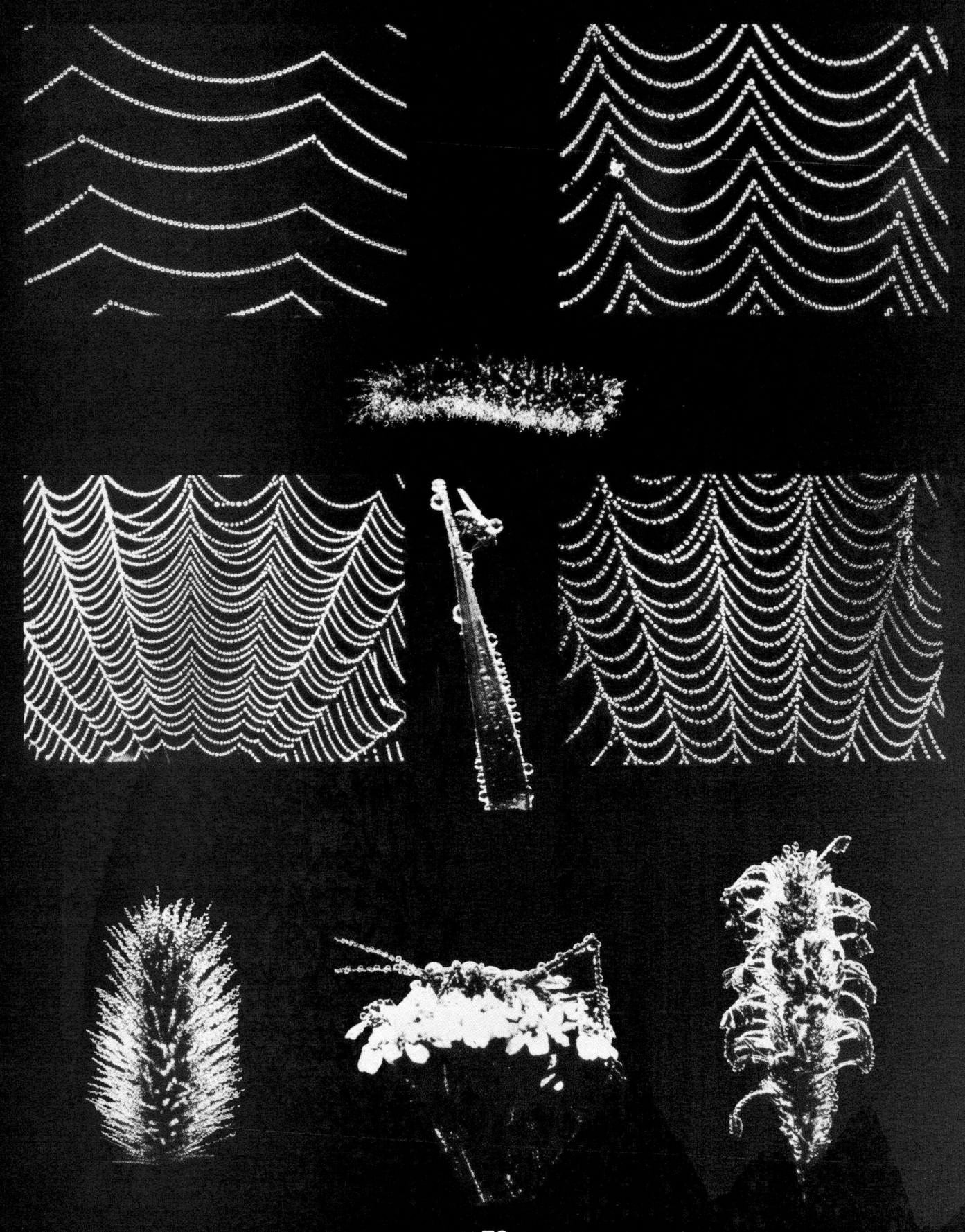